R

SCIENTISM

Can science tell us everything there is to know about reality?

The intellectual and practical successes of science have led some scientists to think that there are no real limits to the competence of science, and no limits to what can be achieved in the name of science. Accordingly, science has no boundaries; it will eventually answer all our problems. This view (and similar views) have been called 'Scientism'.

In this important book scientists' views about science and its relationship to knowledge, ethics and religion are subjected to critical scrutiny. A number of distinguished natural scientists have advocated Scientism in one form or another – Francis Crick, Richard Dawkins, Carl Sagan, and Edward O. Wilson – and their impressive impact both inside and outside the sciences is considered. Clarifying what Scientism is, this book proceeds to evaluate its key claims, expounded in questions such as: Is it the case that science can tell us everything there is to know about reality? Can science tell us how we morally ought to live and what the meaning of life is? Can science in fact be our new religion? Ought we to become 'science believers'? Stenmark addresses these and similar issues, concluding that Scientism is not really science but disguised materialism or naturalism; its advocates fail to see this, not being sufficiently aware that their arguments presuppose the previous acceptance of certain extra-scientific or philosophical beliefs.

Ashgate Science and Religion Series

Science and religion have often been thought to be at loggerheads but much contemporary work in this flourishing interdisciplinary field suggests this is far from the case. The Ashgate Science and Religion Series presents exciting new work to advance interdisciplinary study, research and debate across key themes in science and religion, exploring the philosophical relations between the natural and social sciences on the one hand and religious belief on the other. Contemporary issues in philosophy and theology are debated, as will prevailing cultural assumptions arising from the 'post-modernist' distaste for many forms of reasoning. The series enables leading international authors from a range of different disciplinary perspectives to apply the insights of the various sciences, theology and philosophy and look at the relations between the different disciplines and the rational connections that can be made between them. These accessible, stimulating new contributions to key topics across science and religion will appeal particularly to individual academics and researchers, graduates, postgraduates and upper-undergraduate students.

Scientism

Science, ethics and religion

MIKAEL STENMARK

Ashgate

Aldershot • Burlington • Singapore • Sydney

Published by
Ashgate Publishing Limited
Gower House
Croft Road
Aldershot
Hants GU11 3HR
England

Ashgate Publishing Company
131 Main Street
Burlington, VT 05401–5600, USA

Ashgate website: http://www.ashgate.com

British Library Cataloguing in Publication Data
Stenmark, Mikael
    Scientism: Science, ethics and religion. –
    (Ashgate Science and Religion Series)
    1. Scientism 2. Religion and science 3. Science – Moral and ethical aspects
    I. Title
    215

Library of Congress Cataloguing-in-Publication Data
Stenmark, Mikael
    Scientism: Science, ethics and religion / Mikael Stenmark.
        p.   cm. – (Ashgate Science and Religion Series)
        Includes bibliographical references and index.
        1. Scientism. I. Title. II. Series.
        B67.S825 2001
        501–dc21   2001022839

ISBN 0 7546 0445 4 (hbk)
ISBN 0 7546 0446 2 (pbk)

This book is printed on acid free paper.
Typeset by Owain Hammonds, Bontgoch, Talybont, Ceredigion, Wales SY24 5DP.
Printed and bound in Great Britain by MPG Books Ltd, Bodmin, Cornwall

Contents

Acknowledgements

In writing this book I have incurred debts to many colleagues, friends and institutions. I gratefully acknowledge the financial support of the Swedish Council for Research in the Humanities and Social Sciences which made this whole project possible. I would also like to express my thanks to Philip Hefner for inviting me to the Chicago Center for Religion and Science where I spent six months in 1995 doing research. Vincent Brümmer kindly asked me to give a number of lectures at Utrecht University in 1998. This gave me the chance to formulate these ideas for the first time in public. For this and the thoughtful comments and the encouragement I received I am very grateful. The opportunity to discuss my ideas with Wentzel van Huyssteen and his graduate students at Princeton Theological Seminary in 1999 also helped me improve my arguments. Very important indeed have been my discussions with colleagues and students at the Department of Theology at Uppsala University, especially those with Eberhard Herrmann and Carl-Henric Grenholm. The manuscript was also read in parts by Peter Byrne, Olof Franck and Linda Zagzebski, and I thank them for their very useful suggestions. Most of the material has not been published before, but Chapter 1 previously appeared as the article 'What is Scientism?' in *Religious Studies*, 33 (1997), pp. 15–32 and parts of Chapter 6 as 'Evolution, Purpose and God', in *Ars Disputandi: The Online Journal for Philosophy of Religion*, 1 (2000–2001).

Introduction

Western society has been much shaped by scientific thought and discoveries. We depend not only practically on science in our lives. The theories and methods of science also shape our thinking and attitudes. The overwhelming intellectual and practical successes of science that lie behind its impact on our culture have led some people to believe that there are no real limits to the competence of science, no limits to what can be achieved in the name of science. There is nothing outside the domain of science nor any area of human life to which science cannot successfully be applied. A scientific account of anything and everything constitutes the full story of the universe and its inhabitants. Or, if there are limits to the scientific enterprise, the idea is that science, at least, sets the boundaries for what we humans can ever know about reality. This view (or similar views) has sometimes been called *Scientism*.

Scientism in one version or another has probably been around as long as science has existed. Recently, however, a number of distinguished natural scientists have advocated Scientism in one form or another. The scientists I have in mind in particular are Francis Crick, Richard Dawkins, Stephen Hawking, Carl Sagan and Edward O. Wilson. These scientists, besides receiving a number of prestigious scientific prizes and awards, have sold an enormous number of books. The views of these scholars have been discussed in newspapers and have been broadcast on radio and television. Although Scientism has been around for a while, the great impact these advocates of Scientism have had on popular Western culture is new. They have brought not only science but also Scientism into the living rooms of ordinary people. But, of course, one need not be a scientist to be a defender of Scientism. The view in one version or another is quite popular among philosophers these days.[1] Some politicians can even be viewed as its champions. Jawaharlal Nehru, the first prime minister of independent India, wrote,

> It is science alone that can solve the problems of hunger and poverty, of insanitation and illiteracy, of superstition and deadening custom and tradition, of vast resources running to waste, of a rich country inhabited by starving people. ... Who indeed could afford to ignore science today? At every turn we seek its aid. ... The future belongs to science and to those who make friends with science.
>
> (quoted in Sorell 1991: 2)

The aim of this study is to identify what Scientism is or can be and to critically evaluate the claims made by its defenders; in particular we shall ask whether

---

[1] See, for instance, Churchland (1979) and Dennett (1991; 1995).

Scientism really is science as it pretends to be. Although the term 'Scientism' is frequently used, it is often not clear what it signifies. People have in fact given it a number of different meanings. We must therefore distinguish between these different forms of Scientism. I shall analyse some of these conceptions of Scientism and relate them to one another.

As we shall see, some scientists believe that since we have modern biology, we no longer have to resort to superstition when faced with deep questions such as 'Is there a meaning to life?' 'What are we here for?' 'What is man?' because science is capable of dealing with all these questions and constitutes in addition the only alternative to superstition. Science can be our new religion and answer our existential questions. Indeed, we ought to become 'science believers' and leave our traditional religions or secular ideologies behind. Some forms of Scientism thus offer a substitute for traditional religion and ideology.

Other forms focus on ethics rather than on religion. So we shall examine views which hold that morality is ultimately about selfishness or maximizing fitness. Evolutionary biologists have discovered, we are told, that morality is an illusion fobbed off on us by our selfish genes to get us to cooperate and therefore there is no objectivity to morality and ethics. These scientific discoveries are believed to be the greatest intellectual advance of the twentieth century and to have a profound impact on our self-view; we have to start all over again to describe and understand ourselves, in terms alien to our intuitions. Some of these scientists also believe that contemporary science, with the recent development of evolutionary biology, can even tell us how we morally ought to live and what we ought to value in life. Scientific theory, contrary to what we previously have thought, can justify ethical norms and beliefs and provide us with a new, scientific ethic.

Some advocates of Scientism make less spectacular claims, maintaining merely that science sets the limit for what we can know. All genuine knowledge about reality in general or about human life in particular is to be found through science and science alone. Science is seen by these scientists to be the only begetter of truth. Science is, therefore, taken to be the candle in the dark in a demon-haunted world. It is our only hope to avoid superstition and safeguard our cultural achievements and our planet. Consequently, if there is no truth to be found outside science, scientists must become missionaries and bring the gospel to the pagans and unenlightened people. The broad agenda must be to strive to incorporate many other areas of human life within the sciences, so that rational consideration and acquisition of knowledge can be made possible in these areas as well.

These are exciting claims and if true they would have profound implications for our self-understanding and for the importance we give to the scientific enterprise. So these are some of the issues we face. Can science provide us with ethical guidelines and even replace traditional ethics? Can science tell us what we are here for or what the meaning of life is? How are Scientism, on the one hand, and naturalism or atheism, on the other, related? Can science be one's religion? What is the relationship between Scientism and traditional religions such as Christianity and Judaism? Can science explain traditional religion as a wholly material phenomenon? Or is it still possible to be an intellectually fulfilled religious believer? Is science setting the limits for what can exist and what we can know about what exists? Or are there truths to be found outside of science which are not detectable by scientific methods and experiments? Does science in fact presuppose such truths to function properly? What are the limits of science? What legitimate roles can science play in the development or reconstruction of a world view or religion? I will attempt to give answers or at least partial answers to these questions as we proceed.

I shall suggest that there are good reasons to deny that Scientism is proper science. I shall maintain that many of the claims advocates of Scientism think are scientific statements are in fact not, and these claims cannot, therefore, be made in the name of science. The analysis will show that Scientism typically is a combination of certain scientific theories and a particular ideology or world view, namely, naturalism or materialism; and further, it will be argued that naturalism and materialism are not scientific but philosophical theories. Nevertheless, we ought to take these 'scientistic' claims seriously because science has in the past been able to deal with issues we previously thought it could not deal with. Therefore, the issue about the limits of science cannot be settled once and for all. Instead we need to look again and again at the particular claims of expansion of the scientific domain made by scientists. I am firmly convinced that anyone who wants to be a responsible traditional religious believer or spokesperson for a particular ethical or ideological outlook has to take science seriously and pay attention to the theories developed within contemporary science. For instance, the scientific knowledge that we share common ancestry with all other living things undermines the traditional Western belief that we humans are utterly unique beings. We should, therefore, expect that we can learn some important things about ourselves from studies of the behaviour and genetic make-up of other organisms by evolutionary biologists. That is to say, scientific theories can *in conjunction* with other, non-scientific claims either undermine or confirm ethical, ideological or religious beliefs. But the upshot of this discussion is that scientific theories can seldom do this in a direct or straightforward way. What I shall argue is that we have to find 'a path between an exaggerated view of science's importance ... and an impoverished view of science's importance' (Polkinghorne 1996: xi). I share this view with John Polkinghorne and hope that after reading this book you will do the same. The danger we face at the present time is being forced to choose between two extremes, either radical postmodernism (which makes science into nothing but a social construction) or Scientism. Neither is, I believe, in the end intellectually convincing.

Before explaining how the book is structured, let me say a few words about one limitation of the study. I have chosen to focus my attention not on philosophers (with a few exceptions) but on scientists, especially evolutionary biologists, who either explicitly defend Scientism in one form or another, or presuppose it in their writing. Although philosophers (like myself) think that their arguments typically are more sophisticated and better supported than those of others and that therefore their views should be given the most attention, this is not how the general academic establishment or the public audience sees the matter. It is the scientists' and not the philosophers' views that count. Perhaps I exaggerate a little, but I have in this study chosen to focus almost exclusively on scientists who defend Scientism because of the impact of science on the general public and the authority the opinions of scientists are given even outside their own field of expertise.[2] In doing this I have also tried to write in such a way that no knowledge of philosophy is needed to understand the views and arguments discussed. Nevertheless, I hope to be saying some things that would also interest philosophers and theologians.

The book is arranged as follows. In Chapter 1 I distinguish between a number of different forms of Scientism because a careful inquiry of the way the term 'Scientism' has been used reveals that it is actually given a number of different meanings by its adherents and opponents. Noticing this is important because one can hold on to some scientistic claims but, nevertheless, reject some others. I, therefore, distinguish between academic-internal and academic-external Scientism, and among the external ones, between axiological, epistemic, rationalistic, ontological and existential Scientism.

A careful critical evaluation of all scientistic claims would be too comprehensive for this study, so in the critical examination the focus is narrowed down to four key claims: (1) the only kind of knowledge we can have is scientific knowledge; (2) the only things that exist are the ones science has access to; (3) science alone can answer our moral questions and explain as well as replace traditional ethics; and (4) science alone can answer our existential questions and explain as well as replace traditional religion.

The scientistic claims that the only kind of knowledge we can have is scientific knowledge and that the only things that exist are the ones science has access to are clarified and evaluated in Chapter 2. The focus in Chapters 3 and 4 is on Scientism and morality. The attempt to explain morality by using evolutionary theory is investigated and an alternative non-Darwinian explanation is also proposed in Chapter 3. In Chapter 4 two other ways in

---

[2] See Sorell (1991) for a critical discussion of Scientism in philosophy.

which it is possible for evolutionary theory to be of significance to ethics are considered. These are that evolutionary biology can provide us with new information about human life and its environment that can undermine (or support) existing ethical theories, norms or beliefs and that it can justify ethical norms or beliefs and provide us with a new scientific ethic. In Chapters 5 and 6 we move from ethics to religion, and to questions about the meaning of life. I suggest that in the same way that we characterized three claims about the significance of evolutionary theory for ethics, we can distinguish three claims which hold that evolutionary theory can be of significance for religion. The first is that evolutionary theory can explain the development and maintenance of religion in human life. It can give the best account of why people behave religiously or why they believe in God or a sacred, transcendent reality. This idea is examined and critically assessed in Chapter 5. In Chapter 6, the second and third claims are the principal object of study. We start by examining examples of cases where evolutionary theory is supposed to provide us with new information about human life and its environment that undermines existing religious beliefs such as those found among traditional Christians, Jews or Muslims. We then move on to the astonishing claim that evolutionary theory can replace traditional religions and provide us with a new religion or mythology. In the last chapter some general claims are made about possible motives behind Scientism and a summary is given of the main conclusions reached and of the arguments developed to support these conclusions.

CHAPTER ONE

What is Scientism?

Although the term 'Scientism' is frequently used, it is often not clear what it signifies. People have in fact given it a number of different meanings. We must therefore distinguish between these different forms of Scientism. In this chapter I shall analyse some of these conceptions of Scientism and relate them to one another and to traditional religion.

Scientism within academia

What different meanings has the notion of Scientism been given by its advocates and opponents? One way the term has been used is to refer to a programme or strategy *within* science or academia itself. Hence we could call this version *academic-internal Scientism*. Academic-internal Scientism is the attempt to reduce (or translate) into natural science an academic discipline which has not previously been understood as a natural science, or, if that is not attainable, to deny its scientific status or significance in some way. The defenders of academic-internal Scientism all maintain that the boundaries of natural science can be expanded, in one way or another, into fields of inquiry that have not before been considered parts of natural science. The biologist Edward O. Wilson expresses such a view as follows: 'It may not be too much to say that sociology and the other social sciences, as well as the humanities, are the last branches of biology waiting to be included in the Modern Synthesis' (Wilson 1975: 4).

Sometimes, however, the reduction (or translation) does not stop there, but continues even within the natural sciences themselves. For instance, not only is sociology reduced to biology, but biology is reduced to chemistry, and chemistry to physics.

Both of these forms of Scientism (let us call the former *academic-internal₁ Scientism* and the latter *academic-internal₂ Scientism*) seem to be endorsed by Francis Crick, the co-discoverer of DNA, who writes that 'eventually one may hope to have the whole of biology "explained" in terms of the level below it, and so on right down to the atomic level. ... The knowledge we have already makes it highly unlikely that there is anything that cannot be explained by physics and chemistry' (Crick 1966: 14, 98).

We can perhaps define academic-internal Scientism as:

(1) The view that (a) all, or at least some, of the genuine, non-scientific academic disciplines can eventually be reduced to (or translated into)

science proper, i.e. natural science (*academic-internal₁ Scientism*), and/or that (b) all natural sciences can eventually be reduced to (or translated into) one particular natural science (*academic-internal₂ Scientism*).

The claim is typically not that it is possible right now to accomplish either (a), or both (a) and (b), but that eventually it will be possible. However, its supporters hold that we do at present possess grounds for believing that this goal is likely to be obtained in the future. Tom Settle says it is 'a programme, not yet complete, the explanations only promissory notes in some cases, such as the explanation of mentality by neurophysiology' (Settle 1995: 63).

Methodological Scientism

One common way of interpreting academic-internal Scientism₁ is to understand it as:

(2)   The attempt to extend the use of the methods of natural science to other academic disciplines.

Let us call this version, or similar ones, of academic-internal Scientism₁, *methodological Scientism*. Philip S. Gorski, for example, defines Scientism as 'the attempt to apply the methods of natural science to the study of society' (Gorski 1990: 279). And Tom Sorell writes that it is 'The thought ... that it is highly desirable for the concepts and methodology of established sciences to be spread, and unsatisfactory for, for example, ethics or history to be left in their prescientific state ... [which] captures the Scientism in scientific empiricism' (Sorell 1991: 9).[1]

The problem, however, with this view of Scientism is that it is not really plausible to think that the attempt merely to apply methods of natural science to other academic disciplines would be 'scientistic'. Suppose someone argues for the use of statistics or of inter-subjective procedures (i.e. experimental repeatability) in sociology, and the importance of empirical observations and mathematics in philosophy – does that make her a defender of Scientism? Hardly. We need a stronger requirement than Gorski and Sorell offer to make a claim an example of Scientism. However, if the claim is that *only* statistical (and, for example, no hermeneutical) methods are to be used in sociology, then things are clearly different. Or if the idea is that all proper sociological methods *must* yield a result that can be strictly intersubjectively testable (i.e.

---

[1]   See also Schoeck and Wiggins (1960) for a collections of essays united by this theme. Helmut Schoeck writes in the introduction to this volume that 'the word "Scientism" conventionally describes a type of scholarly trespassing, of pseudo-exactitude, of embracing incongruous models of scientific method and conceptualization' (ix). To some extent Bleicher (1982) can also be understood in this way.

the study must be repeatable in such way that if somebody else carries out the study a second time in exactly the same way the results must be identical), then this idea can be understood as scientistic.

Robert C. Bannister is, therefore, probably correct in classifying a certain view as an expression of Scientism if it contains a claim such as 'a scientific sociology must confine itself to the observable externals of human behavior ... [He continues, saying that this] goal meant an end to the cataloguing of feelings, interests, or wishes as a principal activity of prewar sociologists' (Bannister 1987: 3). Here clearly something more than just the application of some of the methods of natural science is undertaken. What have previously been considered proper objects and methods of sociology are also rejected and replaced. Hence a more appropriate characterization of methodological Scientism follows:

(2') Methodological Scientism is the attempt to extend the use of the methods of natural science to other academic disciplines in such way that they exclude (or marginalize) previously used methods considered central to these disciplines.

Scientism within the broader society

There are, however, other ways of understanding 'Scientism' which may or may not be combined with academic-internal Scientism. What these other forms of Scientism have in common is that they attempt to reduce (or translate) something into science which has not previously been understood as science or, if that is not attainable, to deny its significance. They all maintain that the boundaries of science can be expanded, in one way or another, into non-academic areas of human life (such as art, morality, and religion). They are, therefore, all examples of what I shall call *academic-external Scientism*. We can define this as:

(3) The view that all or, at least, some of the essential non-academic areas of human life can be reduced to (or translated into) science.

Loren R. Graham, in an influential study, has dubbed views similar to academic-external Scientism 'expansionism'. He writes,

> Expansionists cite evidence within the body of scientific theories and findings which can supposedly be used, either directly or indirectly, to support conclusions about sociopolitical [e.g. moral, political, aesthetic, religious] values. The result of these efforts is to expand the boundaries of science in such a way that they include, at least by implication, value questions.
>
> (Graham 1981: 6)

He defines values as 'what people think to be good' (Graham 1981: 4). In my view, however, Graham unnecessarily limits expansionism (or Scientism) to value questions. Hence, the difference between academic-external Scientism and expansionism is that the advocate of the former could but, unlike the latter, need not claim that the boundaries of science can be extended so that it includes *values*. Instead, she could, for instance, claim that all *beliefs* that can be known (or even rationally maintained) must and can be included within the boundaries of science. Science sets the limits for what we possibly can know (or rationally believe) about reality. The only sort of knowledge we can have is the scientific kind of knowledge. So there is a crucial difference between these two concepts that should not be overlooked. Nevertheless, in what follows 'scientific expansionism' will often be used as a synonym for 'Scientism' (or more exactly for 'academic-external Scientism') and someone who advocates Scientism will thus be called a 'scientific expansionist'.

Academic-external Scientism raises the question of whether there exists any domain or practice-external limits of science.[2] Do all the tasks human beings face actually belong to (or are solvable by) science? In its most bold formulation Scientism in this form can be taken to maintain that science has no such limits. We will see that there are also weaker versions of academic-external Scientism which admit that science has some kind of practice-external limits.

Epistemic Scientism

The first and probably most common version of academic-external Scientism (or Scientism within the broader society) that we shall consider consists of the attempt to expand the boundaries of science in such a way that all genuine (in contrast to apparent) knowledge must either be scientific or at least be able to be reduced to scientific knowledge. Ian Barbour defines this view as the claim that 'the scientific method is the only reliable path to knowledge' (Barbour 1990: 4). Roger Trigg writes that Scientism consists of the view that 'science is our only means of access to reality' (Trigg 1993: 90). Michael Peterson *et al.*, offer a third way of explaining this version of Scientism. They write that Scientism is 'the idea that science tells us everything there is to know about what reality consists of ...' (Peterson *et al.*, 1991: 36). We can call this form of Scientism, *epistemic Scientism*, and define it as:

(4)   The view that the only reality that we can know anything about is the one science has access to.

The idea is that what lies beyond the reach of scientists cannot count as knowledge. The only sort of knowledge we have is the scientific kind of

---

[2]   See Rescher (1984) for an excellent discussion of the theoretical limits of science.

knowledge. There are no other valid (non-reducible) epistemic activities apart from science. Carnap seems to express this view when he writes that although 'the total range of life still has many other dimensions outside of science … within its dimension, science meets no barrier. … When we say that scientific knowledge is unlimited, we mean: *there is no question whose answer is in principle unattainable by science*' (Carnap 1967: 290). Two contemporary scientists who explicitly hold this kind of view are Carl Sagan and R. C. Lewontin. In his review of Sagan's *The Demon-Haunted World*, the famous biologist Richard Lewontin agrees with Sagan that people ought 'to accept a social and intellectual apparatus, Science, as the only begetter of truth' (Lewontin 1997: 28).

One question of importance for understanding the merits of Scientism concerns which academic disciplines should be considered scientific ones. What is *science*? Any discipline within academia could, in principle, be called a science. (That is the way the term is typically used in for instance Swedish and German.) What is characteristic of Scientism is that it works with a *narrow* definition of science. Before any reduction or translation has taken place, advocates of Scientism use the notion of science to cover only the natural sciences and perhaps also those areas of the social sciences that are highly similar in methodology to the natural sciences. How broad the definition in the end will be (when the programme is completed) is a matter of how many academic disciplines one thinks could be successfully turned into a natural science.

Thus a claim like 'All knowledge is scientific' should be interpreted to mean that we cannot know anything about reality which is not knowable (either directly or after translation) by the methods of inquiry of the natural sciences. We can also see why this is a reasonable way of understanding Scientism if we consider the most common philosophical criticism of it, namely that a scientistic claim like 'All knowledge is scientific' is not itself a scientific but a philosophical claim and is consequently not itself knowable. (More on this in Chapter 2.) If science were defined by the advocates of Scientism in such a way that philosophy is considered a part of science proper, this criticism would lose its point and, of course, Scientism would also lose its point; it would not be a very controversial view. Such a scenario does justice neither to Scientism nor to its opponents. I am therefore inclined to think that a narrow definition of science is a necessary condition for a view counting as Scientism, and this will be the way I shall understand the concept in what follows.

Rationalistic Scientism

It is not always recognized that it is also possible to maintain a stronger epistemological version of Scientism than the above epistemic one. Epistemic Scientism only denies that any claim or belief that cannot be scientifically knowable can constitute knowledge. We cannot know anything about reality which transcends the limits of science. Now, many people have some religious beliefs. Let us. suppose that their truth cannot be scientifically proven; can these people still be rational in accepting these beliefs? An advocate of epistemic Scientism as defined thus far could accept that. All he is, in fact, claiming is that we cannot *know* whether these beliefs are true. From this proposition alone it does not follow that we are not *rational* in accepting them. What is not scientifically knowable might still be rationally believable.

Nevertheless epistemic Scientism and what I shall style *rationalistic Scientism* are sometimes confused because it is not recognized that knowledge and rationality are two distinct concepts. (Epistemic Scientism could only entail rationalistic Scientism if these two concepts were shown to be identical.) It is, however, fairly easy to see that the conditions for knowledge and for rationality cannot be the same. In general we think that people 2,000 years ago were rational in believing that the earth was flat (their believing satisfying the conditions for rationality), but we would not say that they knew that it was flat (their believing satisfying the conditions for knowledge). If they knew, it follows that the shape of the earth must have changed since then. Hence the conditions for knowledge and rationality cannot be the same.[3] Consequently, one can be rationally entitled to believe things that are not scientifically knowable.

Therefore, a stronger epistemological version of Scientism than epistemic Scientism can be maintained. In fact, Anders Jeffner seems to define Scientism along these lines. He writes that the advocate of Scientism 'accepts as reasons for what one should believe about reality (a) reasons such as those acceptable within empirical natural science and (b) only such reasons' (Jeffner 1978: 46). On such an account science sets not only the limits for what we can know about reality, but also the boundaries for what is rational to believe. We have styled this rationalistic Scientism and can now define it as:

(5)   The view that we are rationally entitled to believe only what can be scientifically justified or what is scientifically knowable.

Bertrand Russell, for instance, betrays the commitment not only to epistemic but also to rationalistic Scientism when he writes that

> God and immortality, the central dogmas of the Christian religion, find no support in science. ... No doubt people will continue to entertain these beliefs, because they are pleasant, just as it is pleasant to think ourselves virtuous and our enemies wicked. But for my part I cannot see any ground

---

[3]   See my book *Rationality in Science, Religion, and Everyday Life: A Critical Evaluation of Four Models of Rationality* (1995a: 216–25) for a detailed discussion of the differences between knowledge and rationality.

for either. I do not pretend to be able to prove that there is no God. I equally cannot prove that Satan is a fiction. The Christian God may exist; so may the Gods of Olympus, or of ancient Egypt, or of Babylon. But no one of these hypotheses is more probable than any other: *they lie outside the region of even probable knowledge, and therefore there is no reason to consider any of them.*

(Russell 1957: 44, emphasis added)

So, according to Russell, central Christian dogmas do not merely fail to be scientifically knowable, there is not even any reason to consider them at all. We are thus not rationally justified in believing them. To show that a (religious) belief is not scientific is, on such an account, sufficient for showing that it is neither knowable nor rationally believable.

Note, however, that defenders of both epistemic Scientism and rationalistic Scientism accept that science has *some* practice-external limits. They can admit that there are other kinds of questions and enterprises besides the scientific one. They can maintain this point because it does not follow from the claim that there can be no knowledge or no rational beliefs in the spheres of life outside science that these other realms are unimportant or less valuable than science. It might be accepted that human beings do not live by knowledge alone, other valid and important human activities exist and are necessary for our flourishing. It might further be accepted that science cannot set the limits for what exists. For example, God or a divine reality might exist. The point according to some is merely that we cannot know (epistemic Scientism) or rationally believe (rationalistic Scientism) anything about such a reality.

Ontological Scientism

Sometimes epistemic Scientism is conflated with yet another form of Scientism. Leslie Stevenson and Henry Byerly maintain that Scientism is 'the view that knowledge obtainable by scientific method exhausts all knowledge ... It is ... to posit that whatever is not mentioned in the theories of science does not exist or has only a subordinate, secondary kind of reality' (Stevenson and Byerly 1995: 212). But one can affirm the view that knowledge obtainable by scientific method exhausts all knowledge and yet deny that whatever is not mentioned in the theories of science does not exist or has only a subordinate, secondary kind of reality. One can do this because epistemic Scientism does not preclude the existence of things that cannot be discovered by scientific investigation or experimentation. If there are such things, all epistemic Scientism says is that we cannot obtain knowledge about them. Epistemic Scientism sets the limits of our knowledge and not the limits of reality.

So a more ambitious form of Scientism, *ontological Scientism*, does not merely state that the only reality that we can know (or can rationally believe) anything about is the one science has access to. It maintains further that only the reality science can discover exists. Hence, Scientism can involve a claim about what kind of things exist 'out there'. We can define ontological Scientism as:

(6)   The view that the only reality that exists is the one science has access to.

Reality is what science says it is. Only entities, causes or processes with which science deals are real, period. If science cannot grasp something, it has no reality. It is only fiction. Ontological Scientism thus entails epistemic Scientism because we could not know anything about what does not exist. We cannot know something about a reality to which science does not have access, because there is simply no such reality.

One way of stating ontological Scientism is to maintain that science has shown that nothing but atoms or material particles exist in the world. This is the idea that the only entities and causes in the world are material objects. Wilson frankly calls this view 'scientific materialism' (Wilson 1978: 201). Carl Sagan also writes, seemingly in the name of science, that

> I am a collection of water, calcium and organic molecules called Carl Sagan. You are a collection of almost identical molecules with a different collective label. But is that all? Is there nothing in here but molecules? Some people find this idea somehow demeaning to human dignity. For myself, I find it elevating that our universe permits the evolution of molecular machines as intricate and subtle as we. But the essence of life is not so much the atoms and simple molecules that make us up as the way in which they are put together.
>
> (Sagan 1980: 105)

Sagan apparently thinks that science has shown us that the only things that exist are material objects and their interactions. We are consequently merely 'molecular machines' which are not essentially different from artifacts (i.e. machines). Sagan further claims that 'the Cosmos is all that is or ever was or ever will be' (Sagan 1980: 1). All this, Sagan thinks, is scientifically knowable, not perhaps when the scientific project will be fully developed or completed, but right here and now. Crick calls these ideas the 'Astonishing Hypothesis':

> The Astonishing Hypothesis is that 'You,' your joys and your sorrows, your memories and your ambitions, your sense of identity and free will, are in fact no more than the behavior of a vast assembly of nerve cells and their associated molecules. As Lewis Carroll's Alice might have phrased it: 'You're nothing but a pack of neutrons.' This hypothesis is so alien to the ideas of most people alive today that it can truly be called astonishing.
>
> (Crick 1994: 3)

The ideas of most people have, according to Crick, unfortunately been shaped by prescientific illusions of religion, but only science in the long run can free us from the superstitions of our ancestors.

In what way are these statements by Crick and Sagan examples of Scientism, that is, of maintaining or assuming ontological (and thereby epistemic) Scientism? Notice first, how in saying these things they expand the boundaries of science. I take it that scientists in general would agree that by using scientific methods we can discover that human beings are made of molecules. But what Crick and Sagan assume is that science can demonstrate that this is all human beings are. The problem is that the premise 'Human beings are made of molecules', does not by itself lead to the conclusion, 'Human beings are nothing but a collection of molecules.' It follows only if one adds the premise, 'What science cannot discover does not exist.'

Thus, the arguments of Crick and Sagan are logically invalid unless they presuppose ontological Scientism. (This would also be true about Sagan's claim that the cosmos is all that is or ever was or ever will be.) Crick and Sagan assume that science does not merely give a *true* but also a *complete* account of human beings. More generally, we could say that scientists adhere to ontological and epistemic Scientism *when they present the scientific account as complete.* No further explanation beyond the scientific one is necessary or even possible.

But if Sagan and Crick, at least, implicitly accept ontological Scientism, then they also by entailment accept epistemic Scientism. Hence by accepting epistemic Scientism only, as I earlier pointed out, it is possible to maintain a weaker position than that of Sagan and Crick. As an advocate of Scientism, one could then argue that the conclusion to draw from the fact that the methods of science can establish that human beings are a collection of water, calcium and organic molecules is not that human beings are nothing other than a collection of water, calcium and organic molecules, but that we cannot know anything else about the nature of human beings. Science sets the limits for what we can know about human nature and to believe anything else about the constitution of human beings is merely superstition or blind faith.

Many people think in the light of statements such as these that Scientism and traditional religions such as Christianity and Islam are necessarily incompatible. John F. Haught says that 'it may not be science but *Scientism* that is the enemy of religion [i.e. theism in this case]' (Haught 1995: 17). Scientism claims that science tells us everything there is to know about reality; it even tells us what can exist and, therefore, religion is seriously undermined or even superfluous. But it is important to note that this is not necessarily the case. Recall that there are forms of Scientism which admit that science has some practice-external limits. They accept that there are other valid questions and enterprises besides science. Hence, if religion is taken to deal essentially with value questions, religion and Scientism (in these forms) can be compatible.[4]

Of course, many believers are not satisfied with such a 'narrow' conception of religion. (Nor is Haught for that matter.) They claim that God really exists and that we can know (or at least are rationally entitled to believe) that God is love, and so on. Is not such a 'broad' conception of religion then incompatible with Scientism? After all, Scientism denies that it is possible to obtain knowledge of God or of a divine reality (epistemic Scientism) and that there exists a transcendent (or non-physical) reality beyond the physical universe (ontological Scientism). But to the contrary, Scientism does not *necessarily* deny these things. While Dawkins, Sagan, Wilson and others think along these lines they could be wrong on *scientific* grounds. This is possible because all Scientism claims is that religious beliefs must satisfy the same conditions as scientific hypotheses to be knowable or rationally believable or to be about something real. Hence, people like Dawkins, Sagan and Wilson take for granted that religious beliefs *cannot* meet these requirements, but this could, of course, be questioned.

Richard Swinburne, among others, argues that theism *can* be confirmed by evidence in much the same way that evidence supports scientific hypotheses. There exist close similarities between religious theories and large-scale scientific theories (Swinburne 1979: 3). Just as science explains phenomena with hypotheses about atoms, genes, forces and so on, theism explains why the universe exists and why it looks the way it looks. Swinburne accordingly writes: 'The structure of a cumulative case for theism was thus, I claimed [in *The Existence of God*], the same as the structure of a cumulative case for any unobservable entity, such as a quark or a neutrino' (Swinburne 1983: 386). Only when one can show that the Swinburnian project (or similar ones) is doomed to fail either because it cannot deliver what it promises or because it misrepresents religious belief (or for some other reason) can one claim that (a rich conception of) theism is incompatible with (epistemic, rationalistic or ontological) Scientism.

Hence, Scientism cannot right away be equated with *scientific naturalism* or *scientific materialism*, given that these are understood as, roughly, the views that matter or physical nature alone is real (all phenomena are merely configurations of matter), and that everything that exists (life, mind, morality, religion and so on) can be completely explained in terms of matter or physical nature. They cannot right away be equated because an advocate of either epistemic, rationalistic or ontological Scientism need not endorse these views. Another way of putting this is to say that scientific materialism or naturalism only leaves open the possibility that God could exist if God is identical with

---

4   See Braithwaite (1971) for a classic defence of this view and Herrmann (1995) for a more contemporary one.

the physical world, and hence closes the door for traditional theism or process theism. We have seen, however, that although perhaps epistemic, rationalistic and ontological Scientism often do have such implications, that is not necessarily so.

Axiological Scientism

Yet another form of Scientism is distinguished by Sorell who defines it as 'the belief that science, especially natural science, is much the most valuable part of human learning' (Sorell 1991: 1). He continues: 'What is crucial to Scientism is not the identification of something as scientific or unscientific but the thought that the scientific is much more valuable than the non-scientific, or the thought that the non-scientific is of negligible value' (Sorell 1991: 9). Gerard Radnitzky understands Scientism in a similar way. He writes that 'the distinction between science and non-science by no means implies that other activities, other realms of life, are less valuable. To draw such a conclusion would be a sure symptom of Scientism, a most unscientific attitude' (Radnitzky 1978: 1011).

The claim Sorell and Radnitzky identify as Scientism is different from the versions of Scientism we have discussed so far in that it has nothing to do with knowledge or ontology directly, but deals instead with value questions. Let us, therefore, call this form of Scientism or any form that deals with values *axiological Scientism*. Sorell and Radnitzky claim that we should define this form of Scientism as something like:

(7)   The view that science is the most valuable part of human learning or culture.

It might be true that it is not a scientific conclusion to say that other realms of life are 'less valuable' as Radnitzky writes, or even 'much more valuable' as Sorell, but is it reasonable to interpret such statements as expressions of Scientism? Suppose one thinks that science is more valuable than art, literature, philosophy, politics or sports. Does that make one's view scientistic? This is true, I would say, only if these and other human activities are of almost no value or, as Sorell also says, of 'negligible' value. Hence it is one thing to claim that science is (much) more valuable than non-scientific realms of human life and another to propose that the non-scientific realms are of very little or no value at all. Sorell and Radnitzky are thus guilty of conflating (a) believing that science should be valued higher than other human activities and (b) believing that non-scientific activities are of little value. Scientism then involves a *depreciation* (or an underestimation as the critics would say) of the non-scientific realms of life. Hence a better way of defining axiological Scientism is:

(7') The view that science is the only truly valuable realm of human life; all
      other realms are of negligible value.

Sorrell also maintains that views of morality like those of Edward O. Wilson
and Michael Ruse are scientistic (Sorell 1991: 166). Ruse's view represents a
form of Scientism because he claims that 'on the basis of [a Darwinian]
factual theory about the nature and process of evolution, you can provide a
total explanation of morality' (Ruse 1998: 256). His basic idea seems to be
that morality is an evolutionary mechanism that promotes the survival of our
genes, no more no less. Wilson writes, 'science may soon be in a position to
investigate the very origin and meaning of human values from which all
ethical pronouncements and much political practice flow' (Wilson 1978: 5).
Not only that, he tells us that through neurophysiological and phylogenetic
reconstructions of the mind, 'a biology of ethics [will be fashioned], which
will make possible the selection of a more deeply understood and enduring
code of moral values' (96). Wilson is not yet quite clear about the content of
the ethics that science will establish, but he is convinced that one day we shall
by means of neurobiology acquire a 'genetically accurate and hence
completely fair code of ethics' (Wilson 1975: 575). Hence, traditional ethics
cannot merely be explained but will eventually be replaced by science.

   If these views are scientistic, they are clearly not examples of axiological
Scientism as Sorell defines it above since they do not claim that science is the
only truly valuable realm of human life. Perhaps Ruse and Wilson also
maintain (7'), but it seems quite possible to defend the form of Scientism
identified here and deny (7'). That is, one could claim that science can fully
explain and provide answers to our moral questions without claiming that the
non-scientific realms are of negligible value. This means that there are at least
two forms of axiological Scientism. Let us call (7') *axiological₁ Scientism* and
the one identified here *axiological₂ Scientism*. We can define the latter in the
following way:

(8) The view that science alone can explain morality and replace traditional
     ethics.

Ethics can be reduced to or translated into science. However, for a claim to be
scientistic in this sense, it must maintain more than that science (the theory of
evolution in this case) is relevant to ethics. Nobody would deny that. It must
rather state that science is the sole, or at least by far the most important, source
for developing a moral theory and explaining moral behaviour. In this case the
appropriate claim is that the morally correct way to conduct one's life is
something that can be derived exclusively from the theory of evolution (or
science, more broadly speaking).

   What is the relation of these forms of Scientism to the ones we have
already identified? For one thing, axiological₂ Scientism does not necessarily

entail epistemic, rationalistic or ontological Scientism. One could claim that science can explain morality completely and replace traditional ethics without maintaining that the only reality we can know anything about is the one science has access to, that we are rationally entitled to believe only what is scientifically knowable, or that the only reality that exists is the one science has access to.

With regard to axiological$_1$ Scientism things are less straightforward. It seems logically possible to claim that science is the only really valuable realm of human life, that other realms are of negligible value, and deny, for instance, that the only reality we can know anything about is the one science has access to. However, if we can have non-scientific knowledge, that would be an argument for thinking that the areas in which we can possibly attain this knowledge are of some (and thus not necessarily of negligible) value. Put another way, the belief that the only kind of knowledge accessible to us is scientific knowledge constitutes reason for thinking that only science is of true importance in human life. This is especially so if one accepts both forms of axiological Scientism. If science can provide both knowledge and values, perhaps we do not have to consider any other realms of life significant.

Existential Scientism

Some scientists seem to have an almost unlimited confidence in science – especially in their own discipline – and about what can be achieved in the name of science. Richard Dawkins says that since we have modern biology, we have 'no longer ... to resort to superstition when faced with the deep problems: Is there a meaning to life? What are we for? What is man?' (Dawkins 1989: 1). According to him science is capable of dealing with all these questions and constitutes in addition the only alternative to superstition. He quotes and agrees with G. G. Simpson, who writes that 'all attempts to answer that question [What is man?] before 1859 are worthless and that we will be better off if we ignore them completely' (Dawkins 1989: 1). What answers does science then give to these questions? Science, Dawkins says, tells us that

> We are machines built by DNA whose purpose is to make more copies of the same DNA ... that is *exactly* what we are for. We are machines for propagating DNA, and the propagation of DNA is a self-sustaining process. It is every living object's sole reason for living.
> (quoted in Poole 1994: 58)

Dawkins here seems to think that evolutionary theory can explain not merely *who* we are, but also *why* we exist and what the *purpose* of our life is. When biology has given us its explanations there is nothing more to add or discover.

Stephen Hawking maintains that scientific cosmological theory will help us answer the question 'why we are here and where we came from. ... And the goal is nothing less than a complete description of the universe we live in' (Hawking 1988: 13). In the end we will, with the tools of science, even 'understand the mind of God' (175).

This could not mean anything less, it seems, than that Dawkins and Hawking think that science is able to offer us salvation, to fulfil the role of religion in our lives. We can and must put our faith in science. This is also the way Mary Midgley understands Scientism. She writes that Scientism is 'the idea of *salvation through science alone*' (Midgley 1992: 37). Science is in 'the business of providing the faith by which people live' (57). We can perhaps call this form of Scientism, *existential* or *redemptive Scientism*, and define it as:

(9)   The view that science alone can explain and replace religion.

Another scientist who expresses a belief in the 'salvific' mission of science is Wilson. He claims that traditional religion will be explained and eventually replaced by science. Wilson posits that science has shown that religious beliefs 'are really enabling mechanisms for survival', and apparently nothing more. Science can explain religion as 'a wholly material phenomenon' (Wilson 1978: 3, 192).

In the place of religion Wilson thinks we should put something he variously styles 'scientific materialism', 'scientific naturalism' or 'scientific humanism' (Wilson 1978: 201, 206). This is, as I mentioned above, the view that matter or physical nature alone is real and that everything that exists (life, mind, morality, religion and so on) can be completely explained in terms of matter or physical nature. He suggests that 'scientific materialism must accommodate them [the mental processes of religious belief] on two levels: as a scientific puzzle of great complexity and interest, and as a source of energies that can be shifted in new directions when scientific materialism itself is accepted as the more powerful mythology' (206–7). Evolutionary theory supplies people with a new scientific myth powerful enough to overcome these destructive consequences of the deterioration of traditional religious myths. However, it is not possible now to predict the form religious life and rituals will take as 'scientific materialism appropriates the mythopoeic energies to its own ends' (206).

Scientific materialism can, therefore, also answer, among other things, our existential questions: it can tell us why we are here, where we come from and where we are going. Since there are really no differences between science and scientific materialism, science can be, and *should* be, our religion. In this form Scientism is thus in competition with traditional religions. Religion cannot only be explained by science, it can also be replaced by science. Existential Scientism is equivalent to scientific materialism or scientific naturalism. (Whether it is science is, of course, another matter. A topic we shall discuss in Chapter 6.)

Comprehensive Scientism

Although I have shown that one can accept a particular form of Scientism without necessarily being committed to the other forms, it is of course possible to accept more or less the whole package. This is also the way Scientism is sometimes understood. Radnitzky maintains that 'Scientism is roughly the view that *science has no boundaries*, i.e. that eventually it will answer all theoretical questions and provide solutions for all our practical problems' (Radnitzky 1978: 1008). Arthur Peacocke writes, 'The tendency of science to imperiousness in our intellectual and cultural life has been dubbed "Scientism" – the attitude that the *only* kind of reliable knowledge is that provided by science, coupled with a conviction that all our personal and social problems are "soluble" by enough science' (Peacocke 1993: 7–8). The important thing to focus on is the last part of Peacocke's statement: science alone can solve all our personal and social problems. All our personal, social, theoretical, practical, moral, existential, psychological (you name it) problems are soluble by science alone.

Perhaps this is also the way we should understand Settle when he writes that the hallmark of Scientism (such as Wilson's scientific materialism) 'is to translate everything into science's terms, as far as it will go – and dump the rest ... [E]verything not within science is to be reduced to science' (Settle 1995: 63).

Let us call this form of Scientism *comprehensive Scientism*, and define it as the view that:

(10) Science alone can and will eventually solve all, or almost all, of our genuine problems.

It is perhaps necessary to give a few comments about the meaning of (10). First, it is not merely that science can solve all these problems. Science now needs no help from any other human practice to do it. It is only science that is able to undertake this task.

Second, comprehensive Scientism in its most ambitious formulation (claiming that science can solve all, and not just almost all, of our problems) contains probably all other forms of Scientism that we have identified. If science alone can deal with all our moral problems, it seems to entail axiological$_2$ Scientism. If it alone can solve all our problems, the other realms of human life seem to be of negligible value (axiological$_1$ Scientism). If only science can give an answer to all our theoretical and practical questions, it seems to embrace epistemic Scientism. It is likely, but not strictly necessary, that comprehensive Scientism would then also include rationalistic Scientism. If science can solve any problem we face, it is probably because what science cannot discover does not exist (ontological Scientism). And since science alone can solve all our problems, the non-scientific academic disciplines must be transformed into natural sciences (academic internal₁ Scientism). The only form of Scientism comprehensive Scientism may or may not include is academic-internal₂ Scientism, that is, the claim that the natural sciences themselves can be reduced to one particular natural science.

Third, the qualification 'eventually' in (10) is important because the claim could hardly be that contemporary science, or even science within a near future, will be able to solve all our problems. Instead it must be what we could call *complete science* (that is, what science will be when the scientific project has been carried through to completion and perfection) that will be able to do that (see Rescher 1984: 3–4).

Finally, a second qualifier is necessary to express appropriately the claim of comprehensive Scientism: science only solves the problems that are 'legitimate' or 'genuine'. There is, as Settle pointed out above, a tendency among the advocates of comprehensive Scientism to dismiss everything that cannot be translated into the terms of science. There is an inclination to deny that those problems are genuine or significant and to claim instead that they are pseudo-problems or unimportant problems.

I have tried to show that Scientism comes in a variety of different forms. We first have to distinguish between Scientism within academia (academic-internal Scientism) and Scientism within the broader society (academic-external Scientism). We made a distinction between two versions of the former (academic internal₁ Scientism and academic internal₂ Scientism). The first is the view that all, or at least some, of the genuine, non-scientific academic disciplines can eventually be reduced to science proper, that is, natural science. To this the second adds that all natural sciences can eventually be reduced to one particular natural science.

Among the versions of academic-external Scientism we identified epistemic Scientism (the view that the only reality that we can know anything about is the one science has access to), rationalistic Scientism (the view that we are rationally entitled to believe only what can be scientifically proven or what is scientifically knowable), ontological Scientism (the view that the only reality that exists is the one science has access to), and existential Scientism (the view that science alone is sufficient for dealing with our existential questions or for creating a world view by which we could live). Further, two forms of axiological Scientism were distinguished. The first claiming that science is the only truly valuable realm of human life, the second that science can completely explain morality and replace traditional ethics.

We have also seen that these different forms of Scientism can be combined in a number of different ways. I called the most ambitious combination 'comprehensive Scientism' because it contains all or almost all of these different forms of Scientism. It claims that science alone can and will eventually solve all, or almost all, of our genuine problems. In short, a narrow definition of science (when science is identified merely with the natural sciences) plus any of the versions (1) to (10) above would turn a claim into Scientism.

This variety of forms of Scientism also shows that we should not equate Scientism with scientific naturalism or scientific materialism because there are other possible forms of Scientism that do not entail an acceptance of scientific materialism or naturalism. This variety among versions of Scientism also demonstrates that the relation between Scientism and a traditional religion such as Christianity is not a given. Only between existential Scientism and traditional religions is there a direct conflict. Other forms of Scientism may be compatible with traditional religions.

The questions we need to consider next are whether Scientism is a reasonable view to hold and in particular whether it really is science.

CHAPTER TWO

The limits of knowledge and reality

Since we have seen that Scientism comes in a variety of different forms, I have decided to scrutinize critically the four versions of Scientism that I found to be most interesting and challenging, namely epistemic Scientism, ontological Scientism, axiological₂ Scientism and existential Scientism. These views can be expressed as four scientistic theses:

T1 The only kind of knowledge we can have is scientific knowledge.
T2 The only things that exist are the ones science can discover.
T3 Science alone can answer our moral questions and explain as well as replace traditional ethics.
T4 Science alone can answer our existential questions and explain as well as replace traditional religion.

In what follows the term 'Scientism' will primarily refer to these theses unless otherwise stated. Let us try to be a bit more explicit about how these four claims are related to each other before subjecting them to criticism.

T1 states that all genuine (in contrast to apparent) knowledge is to be found through science. T2 states not merely that the only reality that we can know anything about is the one science has access to; it maintains further that only the reality science can discover exists. So whereas T1 does not entail T2, T2 entails T1. That is to say, if something cannot be discovered by science, then we cannot know anything about it either. If, for instance, God cannot be discovered by scientific means, it follows that we cannot know anything about God. We can only know something about people's thoughts about God, because these thoughts are, presumably, real, even though the intended object of these beliefs would not exist.

T3 says that science can tell us how we ought morally to behave. In Wilson's case, the idea is that evolutionary theory can be used to obtain moral principles or a genetically fair code of ethics. Hence, traditional moral philosophy is not needed anymore, but can be replaced by a biology of ethics. Wilson writes, 'Scientists and humanists should consider together the possibility that the time has come for ethics to be removed temporarily from the hands of the philosophers and biologized' (Wilson 1975: 562). T4 says that science can tell us who we are, why we exist and what the meaning of our life is. Wilson, as we have seen, does not think that science can give us merely a new creed (i.e. scientific naturalism), but also new myths and perhaps religious rituals (Wilson 1978: Chap. 9). Dawkins, on the other hand, is not ready to go quite as far as this. His idea seems to be that science entails, negatively expressed, atheism or, positively expressed, scientific naturalism, but that it offers no new myths or religious rituals.

Wilson, at least, accepts both of these theses. It is, nevertheless, important to distinguish T3 from T4. It is important because it is quite possible to affirm that evolutionary theory is the sole, or at least the most important, source for developing a moral theory and explaining moral behaviour, but at the same time to deny that biology or any other science can tell us what the meaning of human life is or that it can fulfil the role of religion in our lives. That is to say, one could maintain that evolutionary theory can tell us which ethical principles we should use when trying to solve moral problems concerning, for example, abortion, population growth and conflicts between people of different classes, gender and races, and stop there, thereby accepting that the choice of religion or world view is beyond the scope of science.

Thus T3 does not entail T4. But does T4 entail T3? This is less clear. Religions and world views are in general taken to include some ideas about how we should live and what a good human life is. If this is correct then the acceptance of T4 implies also an acceptance of T3. But, on the other hand, it is perhaps possible to say that science alone can answer *some* of our existential questions and thus that science can *partially* replace religion. In other words, one doubts or denies that science can, so to speak, deliver the whole package in the shape of a complete world view. If this is so, one could maintain, like Dawkins, that every living object's sole reason for living is to be a machine for propagating DNA, but still deny that science can offer ethical guidelines for how we should conduct our life. Science can answer, at least, some of our existential questions, but cannot solve our moral problems. (Of course, some people understand religion purely in moral terms. T4 then seems to collapse into T3.)

What is then the relation of T3 and T4 to the first two theses of Scientism? Well, neither T3 nor T4 entail T1 or T2. It is not inconsistent to claim that science can answer our moral questions and replace traditional ethics or that science can answer our existential questions and replace traditional religion, without maintaining that the only reality we can know anything about or that the only reality that exists is the one science has access to. Although there is, therefore, no logical necessary connection between T3 and T4, on the one hand, and T2 and T1, on the other, these are, nevertheless, often combined. For example the scientific naturalism that scientists like Dawkins and Wilson adopt as their world view or religion is often based on the previous acceptance of T1 or T2. Dawkins says, for instance:

> So where does life come from? What is it? Why are we here? What are we for? What is the meaning of life? There's a conventional wisdom which says that science has nothing to say about such questions. Well all I can say is that if science has nothing to say, it's certain that no other

discipline can say anything at all. But in fact science has a great deal to say about such questions.

(quoted in Poole 1994: 57)

In this passage Dawkins links T1 to T4.

It is worth noticing, however, that it is possible to see T3 and T4 as merely further expansions of T1. On this interpretation, the advocates of T3 and T4 are just more optimistic than traditional proponents of Scientism. What they maintain is that the domain of scientific knowledge can be expanded into completely new areas of human life. Science can or will in the future, besides the empirical knowledge it already delivers, give us *moral knowledge* (knowledge about how we should morally live) and perhaps even *religious* or, if preferred, *existential knowledge* (knowledge about the meaning of life).

To sum up, what advocates of Scientism have in common is that they maintain that the boundaries of science should be expanded to include disciplines (or answers to questions) that have not previously been considered a part of the domain of science. More precisely, theses T1, T2, T3 and T4 state four ways in which it has been maintained that the boundaries of science should be expanded. I suggest that it is nevertheless sufficient that a person accepts one of these four theses to be counted as an adherent of Scientism.

Let us now consider whether Scientism really is science. Is it reasonable to assume that the scientific enterprise can be expanded in such a way that science can tell us (T1) what the limits of knowledge are and (T2) what the limits of reality are? Are these claims that scientists qua scientists can make? This would be the case it seems, *if* T1 and T2 were the product of scientific investigation or experimentation. If so, they would be claims that are (or could eventually be) a proper part of the body of scientific knowledge.

The limits of reality

Let us start our examination with T2, the idea that the only things that exist are the ones which science has access to. Note first that T2 is not always explicitly stated. Rather two of the scientists I gave as an example of this view in the previous chapter, Sagan and Crick, presuppose the truth of T2 in their description of what human beings are. Recall that they write:

> I am a collection of water, calcium and organic molecules called Carl Sagan. You are a collection of almost identical molecules with a different collective label. But is that all? Is there nothing in here but molecules? Some people find this idea somehow demeaning to human dignity. For myself, I find it elevating that our universe permits the evolution of molecular machines as intricate and subtle as we. But the essence of life is not so much the atoms and simple molecules that make us up as the way in which they are put together.
>
> (Sagan 1980: 105)

The Astonishing Hypothesis is that 'You,' your joys and your sorrows, your memories and your ambitions, your sense of identity and free will, are in fact no more than the behavior of a vast assembly of nerve cells and their associated molecules. As Lewis Carroll's Alice might have phrased it: 'You're nothing but a pack of neutrons.' This hypothesis is so alien to the ideas of most people alive today that it can truly be called astonishing.

(Crick 1994: 3)

In what way do Sagan and Crick presuppose the truth of T2 in their description of what human beings are? They do this by assuming that we can, using the methods of physics and biology, establish that human beings are nothing but a collection of molecules or neutrons. The argument is:

(1) Science has discovered that human beings are made of molecules/neutrons and of assemblages of matter reducible to molecules/neutrons, and of nothing else.
(2) What science cannot discover does not exist.
(3) Therefore, human beings are nothing but a collection of molecules/neutrons.

Now we can also see in what way, more precisely, Crick and Sagan expand the boundaries of science. They are not satisfied with claiming as scientists that 'Science can discover things about the character and content of the universe' but want to go further and maintain that 'Only what science can discover about the character and content of the universe is real.' (Let the 'universe' be the name for everything that exists.) The idea is that one could not as a scientist merely claim that we know things about reality (that, for instance, the earth is rotating and human beings are made of molecules), but claim that what science cannot discover does not exist. So if we cannot obtain scientific knowledge about a phenomenon, it shows that it is an illusion.

What should we say about this form of scientific expansionism? Does the project of developing an exhaustive account of reality genuinely fall within the scope of the sciences? I think that the answer is no. But why should claims such as 'Human beings are nothing but a collection of molecules', 'The Cosmos is all that is or ever was or ever will be' and 'We are merely machines built by DNA whose purpose is to make more copies of the same DNA' not be taken to be proper scientific statements?

One reason is that Crick and Sagan overlook a crucial distinction within scientific methodology, namely the difference between making an ontological and a methodological reduction.[1] As a result they end up making claims about reality that go beyond what the methods of science can establish. For example, and to simplify a bit, what a biologist who examines human beings

---

[1] Both Barbour (1990: 4–5) and Haught (1995: 73–4) make a similar distinction. See also Stenmark (1995a: 178–9).

with her instruments sees is a collection of molecules. Thus, the biologist's instruments warrant her in concluding that humans consist of a collection of molecules. In a methodological reduction she would maintain that for a particular purpose she will reduce a living human being, *A*, to a set of molecules, *B*, but without claiming that *B* is all the human being is. However, we are inclined to think that these instruments are insufficient to determine every aspect of what it means to be a human. To do that it is necessary to use other methods or perspectives. When employing these perspectives we come to think that, besides being a collection of molecules, humans are also beings with self-awareness, an ability to love and so forth. Thus, the biologist's instruments give a true, but not necessarily a complete understanding of what it means to be a human being. The idea conveyed by this illustration is that the scientist is like a fisherman casting a net in the water. Even when the net is used in an optimal way, it does not rule out that there are fish smaller than the mesh of the net or that there can exist other living things in the water.

What Crick and Sagan are doing is something that goes beyond this. Crick and Sagan are in fact assuming that the instruments they use demonstrate or will eventually demonstrate not only that humans are a collection of molecules (or a pack of neutrons) but that they are *nothing more* than that. Crick and Sagan are making an *ontological reduction* because they maintain that a human being, *A*, is nothing more than a collection of molecules, *B*. They claim that a human being, *A*, can be exhaustively described and explained in terms of a collection of molecules, *B*. What cannot be caught in their scientific net or by their instruments does not exist or at least is not significant.

But given that the idea that scientists are only allowed to make methodological reductions is the 'official' scientific doctrine, why could an ontological reduction not be scientifically acceptable? Perhaps Crick and Sagan are proposing a change of scientific methodology, which they for good reasons think is an improvement over previous practice. The problem with this interpretation is, first, that they do not offer any reasons at all for why ontological reductions of this kind should be accepted by the scientific community. They simply *assume* that the scientific account of human beings is complete. Crick and Sagan do this presumably because they believe that all knowledge is to be found through science, and perhaps Haught is right, that if you believe this then you will also believe that matter alone is real, since matter is what science can know something about (Haught 1995: 80). This is, as I pointed out earlier, an invalid inference but it is nevertheless very likely that the key issue really concerns the plausibility of T1, and to this topic we will turn in the next section.

Second, how do you set up a scientific experiment to demonstrate that science or a particular scientific method gives an exhaustive account of reality? I cannot see how this could be done in a non-question begging way. What we want to know is whether science sets the limits for reality. The problem is that since we can only obtain knowledge about reality by means of scientific methods (that is T1), we must use those methods whose scope is in question to determine the scope of these very same methods. If we used *non-scientific* methods we could never come to *know* the answer to our question, because there is according to scientistic faith no knowledge outside science. We are therefore forced to admit either that we cannot avoid arguing in a circle or that the acceptance of T2 is a matter of superstition or blind faith.

Furthermore, if, for instance, God (or a divine reality) exists, it seems almost uninformed to expect that such a reality should be graspable by means of scientific investigation and experimentation. If God exists, we would expect God not to be that kind of being (or most religious believers would think so anyway). To put it bluntly, scientists would never be able to see God through their telescopes or microscopes or by using linear accelerators or cloud chambers. And this is not surprising. But there are also other things which many of us believe exist that cannot be discovered in such ways. We can certainly scientifically describe a painting as a canvas of a particular dimension containing certain pigments with a thickness of 5 mm. We can also describe it in terms of the quality of the light, the motion of the figures and so forth. However, the depth of painting we discover in this way cannot be reduced merely to the thickness of the pigments. By using the methods of science I cannot determine that the depth of the painting is *nothing more* than the pigments with a thickness of 5 mm. To do so is to unwarrantedly assume that the only valid perspective of reality is the scientific perspective.

Hence, science cannot exclude the possibility that there are dimensions of reality which are neither describable in scientific language nor accessible to scientific explanations, simply because that issue is beyond the competence of science. We must also remember that the issue is not whether one can claim that science gives an exhaustive account of reality, but whether that claim is a *scientific* claim. But it cannot be a scientific claim since it cannot be evaluated by means of scientific investigation and experimentation. Rather it must be an *extra-scientific* claim, a *philosophical* claim, and when a claim such as this is added to science, you typically end up with a particular ideology, namely scientific materialism. Although this is a philosophically acceptable view, it should not be conflated with science.

Thus, Crick's claim 'We are nothing but packs of neutrons', Sagan's 'The Cosmos is all that is or ever was or ever will be' and Dawkins' 'Every living object's sole reason for living is that of being a machine for propagating DNA', are extra-scientific or philosophical claims. Because Crick, Dawkins and Sagan do not especially distinguish these claims from scientific ones, they are using science in an ideological way. Thus, Crick, Dawkins and Sagan should inform their readers that they are not only explaining and defending certain scientific theories but also advocating a particular ideology, a reductionist scientific naturalism.

Perhaps it is true that it is not possible to scientifically justify T2 and that it is a philosophical claim, but is there really any reason to think that there is more to reality than what the sciences can discover and furthermore that we can know something about these dimensions of reality? I have already suggested that that is the case, but let us try to answer this question by examining T1, because if T1 turns out to be false then T2 cannot possibly be true.

The limits of knowledge

Scientists who accept T1 believe that all genuine knowledge about reality is to be found through science and science alone. Hence although there could be more to reality than what we can discover by employing the methods of science, we surely cannot know anything about it. Everything outside science is therefore typically taken as a matter of mere belief and subjective opinions. Consequently, the agenda of the scientists who accept T1 is to strive to incorporate many other areas of human life within the sciences, so that rational consideration and acquisition of knowledge can be made possible in these fields as well. This is a part of the mission of the scientistic faith. In a 'demon-haunted world', science is the 'candle in the dark' (Sagan 1997).

Just as with T2, T1 is not always explicitly stated by its scientific advocates. But of course any scientific reasoning that presupposes the acceptance of T2 also presupposes T1, since T2 entails T1. Let us, however, go back to one of the passages quoted in Chapter 1 to illustrate this view. Dawkins writes that since we have evolutionary theory,

> We no longer have to resort to superstition when faced with the deep problems: Is there a meaning to life? What are we for? What is man? After posing the last of these questions, the eminent zoologist G. G. Simpson put it thus: 'The point I want to make now is that all attempts to answer that question before 1859 are worthless and that we will be better off if we ignore them completely.'
>
> (Dawkins 1989: 1)

Dawkins laments in the second edition of *The Selfish Gene* that even non-religious people have questioned these claims. But he stands firm and cannot think of any pre-Darwinian claim about these matters that 'are not now worthless except for their (considerable) historical interest' (Dawkins 1989: 267).

It is reasonable to assume that Dawkins and Simpson in telling us these things simply presuppose the truth of T1. Only if they take for granted that genuine knowledge can be obtained solely by employing the methods of science, does it seems reasonable to state in this *a priori* fashion that no other discipline can say anything at all about these questions and that all human inquiries of human nature before the development of Darwinism are simply worthless. (One cannot help but wonder whether they have ever read anything of, for example, Plato or Aristotle about human nature and our predicament!)

The expansion of the boundaries of science that T1 envisages consists of the move from accepting that 'Science gives us knowledge of reality', to maintaining that 'Nothing but science gives us knowledge of reality.' How are we to evaluate this form of scientific expansionism? Does the project of determining the limits of knowledge actually fall within the scope of the sciences? Notice that there are really two questions we must address, namely:

(1) Is scientific knowledge the only kind of knowledge we can have or is it only one particular sort of knowledge?
(2) Can question (1) be answered by the sciences, that is, by scientific investigation and experimentation?

Question (2) is crucial. We cannot *know* that scientific knowledge is the only mode of knowledge unless we are able to determine this by scientific means. This is so, simply because science sets the limits for what we can possibly know. The option left is to consider these scientists' belief – that science sets the limits of our knowledge – as merely an expression of blind faith or of a subjective opinion.

Let us start with question (1). One way of assessing T1 is to consider whether there are things we know which are not scientifically knowable. If that is the case then there would be other valid epistemological activities apart from the scientific one, and T1 would have been shown to be false.

Clearly a number of people think that such forms of knowledge exist. Alan Chalmers, for example, writes,

> In addition to what is typically regarded as scientific knowledge, we have everyday, common-sense knowledge, we have the knowledge possessed by skilled craftsmen or wise politicians, the knowledge contained in encyclopaedias or stored in the mind of a quiz show expert, and so on.
>
> (Chalmers 1990: 25)

Furthermore, Mary Midgley maintains,

> Science cannot stand alone. We cannot believe its propositions without first believing in a great many other startling things, such as the existence of the external world, the reliability of our senses, memory and informants, and the validity of logic. If we do believe in these things, we already have a world far wider than that of science.
>
> (Midgley 1992: 108)

I think this is quite convincing as it stands, but it would nevertheless be desirable if we could be a bit more specific and give some concrete examples of instances of these non-scientific modes of knowledge. Is this possible? Consider the following beliefs that I have: 'I exist', 'I am now thinking about Scientism', 'There is a pain in my stomach', 'I am in love', 'I can trust my memory' and 'What is written above is a statement by Chalmers about different kinds of knowledge.' Furthermore, by looking out through my window, I come to believe that I see a tree and that people are walking on the street and that my friend John is standing there on the street and greets me by raising his hand. Are these beliefs scientific ones or reducible to such beliefs and if not, is it reasonable to think that these beliefs can constitute knowledge?

Let us start with what most obviously looks like scientific belief, namely the belief that there is a tree outside my window and that people are walking on the street. These are just two examples of the kind of *observational knowledge* we acquire every day. One difficulty with classifying observational knowledge as scientific knowledge is that it would mean that we are all scientists and further that 'science' has existed long before the development of science. We all (or almost all of us since some people are blind) must be scientists since we have the means of acquiring these beliefs without seeking expert advice, that is, by perception alone we can come to know 'There is a tree outside my window' and 'People are walking on the street.'

Compare these beliefs with what I would classify as scientific beliefs, such as 'Genes are segments of chromosomes', 'Chromosomes are composed of DNA', 'Nuclear fusion causes the sun's energy' and 'All particles of light travel with a velocity of 300,000 km/sec.' These beliefs, in contrast to observational beliefs, are obtained by means of scientific methods and experimentation. Scientific knowledge presupposes the development of methods and experimental techniques and we are not all scientists since most of us do not master such methods and techniques. Thus, science aims to understand the constitution and causes of the physical world and seeks to give an account of what is going on behind the phenomena we experience, that is, what the world is like in the realm that is too small, too distant or too far in the past to be directly experienced. This is done by developing theories about, for instance, the transmission of diseases, the motions of planets and stars, the succession of fossils and the similarities among organisms.

This difference between observational and scientific knowledge can also be expressed by distinguishing between *direct knowledge* and *indirect knowledge*. Right now I see a person passing by outside my window. My knowledge of this would be an example of direct knowledge. Suppose instead that I see footprints in the snow outside my window. My knowledge that a person has passed by my window would, under these circumstances, be an example of indirect knowledge. My knowledge in the second case is indirect because it is inferred from other beliefs I have (in particular the belief 'I see footprints'). Phrased differently, the assumption that a person has passed by my window provides the best explanation of the observed data, that is, of the footprints. In the first case, however, I simply see a person passing by. I neither see something else (like the footprints) from which I infer this person nor offer a best explanation. This also explains why we find it inappropriate or puzzling to talk about theories or hypotheses when dealing with observational beliefs. But it is, of course, not inappropriate or puzzling to talk about theories or hypotheses in science. This is so, simply because scientific knowledge is characteristically a species of indirect knowledge.

Hence observational knowledge is not scientific knowledge. Observational beliefs and knowledge are rather things that science typically takes for granted. Science starts from these things. Consequently, if scientific knowledge is the only sort of knowledge we can have, then science itself seems to be based on blind faith or superstition!

I suppose, however, that at least some advocates of Scientism would grant me this point and revise their original claim. They would then maintain that the only things we can obtain knowledge about are the ones that are directly observable or accessible to the methods of science. Let us, therefore, ask whether there are any other things we can know, besides things we can directly observe, that do not constitute scientific knowledge? What about my beliefs 'I am now thinking about Scientism', 'There is a slight pain in my stomach' and 'I am in love'? Are they scientifically knowable? And if they are not, cannot I (or people who have experienced similar things) have knowledge about these things? To be honest, I cannot see how any of these beliefs could constitute scientific knowledge. By using scientific methods (say by measuring my brain waves), scientists can perhaps determine whether or not I am thinking, but they cannot discover what I am thinking *about*, the *content* of my thoughts.

Furthermore, these beliefs of mine are much better justified than any scientific belief or theory. No one can say sincerely, 'There is a pain in my stomach' and be mistaken. One can be mistaken about the intensity of a certain pain but not about the feeling of pain. This is true not only about feelings but also about thoughts. It makes no sense to question my belief that I am now thinking about Scientism by asking, 'Mikael, are you certain about this? Is it not rather the case that you are thinking about food?' (This is so, unless, of course, you are questioning my truthfulness, thinking that I am trying to fool you.) Thus, we have good reasons to believe that besides scientific knowledge and observational knowledge we also have *introspective knowledge*.

Many people have also thought that since we believe that we have feelings and thoughts, it must be that these things *exist* and further that we can *know* that this is so. But because thoughts and feelings are in the mind, it seems as if at least some things that exist in the world are mental or non-physical things. Most of us, I assume, would on reflection also admit that we are more certain about this than about the existence of some physical things. That is to say, we do not have grounds for claiming that non-mental things exist that are as solid as our grounds for being certain about the contents of our own minds. While one can certainly be wrong about things outside oneself, one cannot be wrong (or so it seems) about one's feelings and thoughts. In saying this I by no means intend to deny that there exist objects that are non-mental and claim that everything that exists occurs within the mind or minds (the view philosophers call 'idealism'). All I am saying is that it seems as if the beliefs we have about our inner life are more likely to be true than those about the external world. Thus introspective knowledge is more certain than observational knowledge and scientific knowledge.

We have knowledge, or at least so it seems, about non-physical things such as our feelings and thoughts. But what about the belief that *we* (you and me) are the ones who have this knowledge about feelings and thoughts and, for that matter, about cars and trees? Is the belief that *I* exist – and thus am the locus of these feelings and thoughts – a scientific belief, and if not, can it still constitute knowledge? We have experiences of personal identity over time. We are aware that we are individual centres of consciousness, who now exist and also existed ten years ago. With us matter has become aware of itself. It is furthermore very difficult to doubt that I exist, and am the one who is feeling a pain in the stomach and who is now thinking about Scientism. It is hard to question that I *know* that *I* exist. Is this belief that there is an 'I' or a 'self' a scientific belief? No, I certainly have neither acquired nor tested that belief by using scientific methods. In fact it seems very hard to do such a thing. Must I not assume my own existence at the outset of such a scientific inquiry, in order to believe that I was the one collecting and examining the relevant evidence? The answer must be yes, thus making the whole process hopelessly circular. I would have to assume at the outset, what I intended to prove (or disprove).

Moreover, only a self-conscious being knows that he is a knower, that is, someone who is capable of being aware that he knows certain things. One cannot have *self-reflective knowledge* if one does not know that it is oneself who has this knowledge. Hence, the advocates of Scientism can only consciously know that the proposition 'Scientific knowledge is the only kind of knowledge we can have' is true, if they also know that *they* are the ones who know this. But then they have to admit that they know at least one thing that is not scientifically knowable (namely that they have a self). So if they know that they know that the only things we can obtain knowledge about are the ones science has access to, they cannot know it because it is false. This point holds with respect to all self-reflective knowledge we have (and that is a lot!), for example my knowledge that *I* am writing this paper on Scientism, *I* am married to Anna, *I* am looking at the tree outside my window and that *I* am a philosopher of religion.

We already have very good reasons to reject T1, but let me give some more examples of non-scientific sorts of knowledge because it is important in an age that is so dominated by science that we realize just how much knowledge we have that is not scientific. Let us therefore turn to yet another group of beliefs, namely *beliefs of memory*. That is those beliefs which are about things we have previously experienced or thought about. For instance, I remember that I am married to Anna and fell in love with her in 1986, and that I am writing about Scientism. Furthermore, I do not merely believe these things, I also think that I know these things. But I do not think that the beliefs of memory can be scientifically proven. Rather, to be able to develop and test a scientific hypothesis against a certain range of data, scientists have to be able to remember, for instance, the content of hypothesis, the previous test results and more fundamentally that they are scientists and where their laboratories are located. Their scientific knowledge presupposes memory.

The truth is that unless we could trust our memories (and obtain knowledge), we could never reason at all or do any science whatsoever, because in any inference we must remember our premises on our way to the conclusion. All activities we are engaged in therefore presuppose knowledge based upon memory. Accordingly, science not merely takes observational knowledge and self-reflective knowledge for granted, it also presupposes the possibility and reliability of knowledge based upon memory. But if scientific knowledge is the only sort of knowledge we can have, then we cannot know that we know this because such knowledge requires knowledge based upon memory.

Let us once more go back to the kind of beliefs listed at the beginning of this section. What about my belief 'Chalmers writes about different kinds of knowledge' and my belief that on the billboard sign on the street it says, 'DRINK COCA-COLA'. By reading the quotation by Chalmers above I obtain the knowledge that he writes about different kinds of knowledge. Similarly, by looking at the billboard sign I become aware of the fact that the words 'Drink Coca-Cola' appear on it, and I *know* what those signs mean. Now, to put it bluntly, can science read books or, for simplicity, these two sentences,

'Chalmers writes about different kinds of knowledge'
'Drink Coca-Cola'

and thus obtain *linguistic knowledge*?

More precisely, the question is whether, for example, the biologist *qua* biologist or the physicist *qua* physicist can read these texts. Can they as scientists discover (or come to know) the meaning of these sentences by applying solely the methods of biology or physics? Well, scientists can, of course, analyse the chemical laws that allow ink to bond with paper and the other things that make it possible to write these sentences. But can scientists with these methods come to know the information contained in these sentences? I must admit that I cannot even imagine what such an experiment would look like.

The crux of the matter is that it is not even possible to become a scientist without first being a 'hermeneutic creature', that is, a being that can understand and interpret meaningful phenomena (that is, things which express a meaning) such as languages. Phenomena like languages, works of art, scientific theories, music and religious stories belong to what Karl Popper refers to as World 3, the world of cultural products (Popper and Eccles 1977: Chap. 2). World 1 is the universe of physical things and World 2 is the world of mental things. The two sentences above are physical things, and they therefore belong to World 1, but they are not merely physical things because they have a content and this content – their meaning – belongs to World 3 (a content which was originally World 2 thoughts). Hence, these sentences are not, as Crick seems to suggest, nothing but a pack of neutrons. That is to completely miss the most important thing they are, namely that they are meaning carrying signs. Turning to scientific theories, suppose we ask if it is true that the theory of evolution is nothing but a pack of neutrons? Hardly! Furthermore, World 3 things are real because they can affect World 1 and World 2 events. For example, the content of the recipe affects the way I bake the cake and the information conveyed on the street map affects the route the tourists decide to take, to see what they came to see in Uppsala.

In Popper's terminology, linguistic knowledge is thus knowledge about certain World 3 things and the problem for advocates of Scientism is that although the focus of biologists and physicists is on World 1, *they must develop and use World 3 things such as languages and theories as their tools in order to obtain knowledge about World 1*. Therefore, if we cannot have genuine linguistic knowledge or knowledge about at least some meaningful things, we cannot successfully develop a scientific practice. Science takes *hermeneutics* for granted. Consequently, not only does it seem as if science presupposes World 2 things (selves and their mental states) and World 2 knowledge (self-reflective knowledge and knowledge based upon memory), but it also presupposes World 3 things (languages and theories) and World 3 knowledge (linguistic knowledge).[2]

We are not only able to know the meaning of the sentence 'Drink Coca-Cola' written on the billboard, but we are also able to know that someone put it there with the intention of persuading us to buy a certain type of product. There is a reason for its being there and we can know this. Thus, we have not merely linguistic knowledge but also *intentional knowledge* (that is,

---

[2] Furthermore, and as Popper points out, we will never be able to understand the behaviour of scientists unless we admit that meaningful phenomena such as theories are real, knowable and thus possible objects of study and of criticism (Popper and Eccles 1977: 40).

knowledge about people's intentions or purposes) or so it seems. A scientist can, of course, explain the lighted billboard in terms of the strength of steel posts that hold the sign, the movement of electricity which causes the lights to glow and so forth. But meaningful phenomena such as intentions are clearly nothing that scientists can obtain knowledge about by merely applying the methods and instruments of physics or biology or any other natural science for that matter.

The advocates of Scientism therefore are forced to deny that there are such things as intentions or purposes and, consequently, maintain that our purported knowledge about them is merely an illusion. In this particular case, they have to deny that the lighted billboard contains a dimension of reality that is undetectable by their scientific methods, namely, that the sign expresses an intention to persuade us to buy a certain type of product (which would be in conflict with T2) and, furthermore, that we can obtain reliable knowledge about this (which would be in conflict with T1). But is it really reasonable to deny that intentions exist and that we can have intentional knowledge and, more importantly, can they *as* scientists deny this?

Can we even understand science if we deny the possibility of intentional knowledge and thus intentional or teleological explanations? If in trying to explain why Darwin developed the theory of evolution, for instance, we were merely to refer to the molecular movement in his brain or the propagation of DNA, would we have fully understood his behaviour? The answer it seems must be no. No satisfying account of Darwin's behaviour can avoid referring to Darwin's *intention* to explain the diversity of living things and trace the patterns in that diversity and his *belief* that his theory offers the best explanation of these phenomena. Darwin's intention to explain the diversity of living things is obviously not the same event as some nerve firing in the brain. More importantly, the content of his thoughts (including his intentions) cannot be discovered by scientifically examining some nerve firing in the brain. Nor, for that matter, can the content of the theory of evolution be discovered in such a way.[3]

One move advocates of Scientism can make at this point is to maintain that even though intentions may exist, they (and other mental states) really make no difference to what happens. They are merely epiphenomena like our own shadow. They are not causally effective. But, as Richard Swinburne points out, this goes strongly against our personal knowledge. 'We know very well that, if we ceased to form purposes and to try to execute them, nothing

---

[3] Some philosophers have developed theories which deny the existence of intentions and beliefs, see, for instance, Churchland (1979) and Stich (1985). But I do not have any problem with them, *as philosophers*, developing and defending these theories, because they are not doing anything that falls outside the scope of philosophy. I think, however, that they are wrong. For a critique of these views see Haack (1993: Chap. 8).

would happen' (Swinburne 1996: 22). If I did not intend to finish this sentence, I would not finish it, nor would I walk or talk. What we are trying to achieve makes all the difference to what happens in these situations.

Hence, vast stretches of our cognitive landscape consist of knowledge about people's intentions and in both everyday life and the humanities we frequently have to use teleological explanations to make sense of our own and other people's behaviour. Different events or phenomena are explained in different ways, and the outcome of these explanations, if successful, yields different types of knowledge.

The most troublesome difficulty with T1, however, is that it appears to be *self-refuting*, that is, T1 seems to tell us not to accept T1. This is a very serious problem for the defenders of Scientism, because if T1 is self-refuting then it is not even possible for T1 to be true. T1 would be necessarily false in the same way as it is logically impossible for the propositions 'Some married men are bachelors', 'There are no truths' and 'There is a human being who is taller than himself' to be true and to constitute knowledge. So no matter how successful science turns out to be in the future, it will not make the slightest difference for the truth of T1.

What reasons are there for thinking that T1 might be self-refuting? Recall that in order to assess T1 we had to answer two questions, 'Is scientific knowledge the only kind of knowledge we can have or is it only one particular sort of knowledge?' and 'Can the first question be answered by the sciences, that is, by scientific investigation and experimentation?' The second question is crucial because we cannot *know* that scientific knowledge is the only mode of knowledge unless we are able to determine this by scientific means. This is so, simply because T1 tells us that science sets the limits for what we can know. The only other option left for the advocates of Scientism is to see T1 as an expression of blind faith, of a subjective opinion or perhaps of superstition, which, of course, is something they want to avoid at any cost.

The problem is that the scientistic belief that we can only know what science can tell us, seems to be something that science cannot tell us. How can one set up a scientific experiment to demonstrate the truth of T1? What methods in, for instance, biology or physics are suitable for such a task? Well, hardly those methods that make it possible for scientists to discover and explain electrons, protons, genes, survival mechanisms and natural selection. Furthermore it is not because the content of this belief is too small, too distant or too far in the past for science to determine its truth-value (or probability). Rather it is that beliefs of this sort are not subject to scientific inquiry. We cannot come to know T1 by appeal to science alone. T1 is rather a view in the theory of knowledge and is, therefore, a piece of philosophy and not a piece of science. But if this is the case, then T1 is self-refuting. If T1 is true, then it is false. T1 falsifies itself.

Hence, if the advocates of Scientism do not want to see T1 as an expression of blind faith, of a subjective opinion or perhaps of superstition, their only option is to revise T1 in such a way that it is not self-refuting. This can be achieved if the thesis is given the following content:

(T1´) Scientific knowledge is the only kind of knowledge we can have with the exception of one philosophical claim, namely this claim, which we can also know to be true.

The problem with this move is, of course, to give a convincing argument for why this particular claim and no other philosophical claims can be known to be true. But in doing this one cannot any longer rely on the authority of science. The game is now to be played in the philosophical courtyard. Furthermore we have already seen that there are actually other sorts of knowledge we can obtain besides scientific knowledge, so there are good reasons to reject not only T1 but also T1´.

In this chapter the scientistic theses T1 and T2 have been subjected to criticism. In addition to the embarrassing fact that the claim 'We can only know what science can tell us' is self-refuting, we have seen that it is reasonable to believe that our cognitive universe contains much more than scientific knowledge. There are domains of knowledge outside and independent of science. We have at least the domains of observation, introspection, self-reflection, memory, language and intention. Scientific knowledge, furthermore, presupposes the existence of other reliable sorts of knowledge such as that derived from self-reflection, memory and language. So if we did not have these other kinds of knowledge, we would not be able to obtain any scientific knowledge at all. T1 and T2 are also shown to be philosophical rather than scientific claims. Therefore, our conclusion must be that when scientists such as Crick, Dawkins, Sagan, Simpson and Wilson make these scientistic declarations, they are not speaking as scientists and are not doing proper science.

CHAPTER THREE

The scientific explanation of morality

The principal achievement of biological science in the twentieth century was its unfolding of how the secret of life is coded into the genes. This achievement in genetic and molecular biology has been coupled with Darwinian evolutionary biology resulting in a synthesis with great explanatory power. A number of scientists have recently argued that this new Darwinian synthesis has much greater scope of application than many have previously thought. They are ready to apply evolutionary theory to all aspects of human existence and to develop a new Darwinian social and human science, holding that evolutionary theory can yield profound consequences for our understanding of human behaviour and institutions, and maintaining that survival and reproduction are the basic determinants in all human affairs.

This research programme has been named 'sociobiology', although some of its practitioners avoid the label, preferring to see themselves as behavioural ecologists, Darwinian anthropologists or evolutionary psychologists. Edward O. Wilson defines sociobiology as 'the systematic study of the biological basis of all social behavior' (Wilson 1975: 4). Since, for instance, science itself, politics, morality and religion are forms of social behaviour, they fall within the scope of sociobiology. These phenomena can thus be explained by evolutionary theory.

Sociobiology is an example of Scientism (or scientific expansionism) in at least two ways. Central to the research programme are the ideas that biological principles can be extended (1) to the social sciences and the humanities (explaining a new kind of fact) and beyond that (2) into the domains of morality or moral reasoning (explaining and justifying norms and values).

First, evolutionary theory is taken to explain not merely natural phenomena but also cultural phenomena. The claim is that biological explanations of human behaviour can replace the traditional explanations used within the social sciences and humanities. Wilson thus thinks, as we have seen, that it 'may not be too much to say that sociology and the other social sciences, as well as the humanities, are the last branches of biology waiting to be included in the Modern Synthesis' (Wilson 1975: 4).

Second, evolutionary theory is taken to be able to explain morality and to tell us how we morally ought to behave. It can provide us with a new set of values and an evolutionary ethic will be fashioned. Wilson thinks biologists will one day discover a 'genetically accurate and hence completely fair code of ethics' (Wilson 1975: 575). To be able to achieve this goal philosophers ought to be given a long sabbatical: 'Scientists and humanists should consider together the possibility that the time has come for ethics to be removed temporarily from the hands of the philosophers and biologicized' (562). Evolutionary biology can be not merely a source of facts but also a source of norms and it can replace traditional ethics. In fact, he thinks that in 'no other domain of the humanities [than philosophy] is a union with the natural sciences more urgently needed' (Wilson 1998: 62). It is needed because 'it is astonishing that the study of ethics has advanced so little since the nineteenth century. The most distinguishing and vital qualities of the human species remains a blank space on the scientific map' (62).

The sociobiological research programme is thus a form of scientific expansionism or Scientism. Science (in particular biology) really has the answers not merely about human nature and the content of the universe, but about how we ought to live. What I shall then focus on in this and the succeeding chapter is the thesis T3, that is, the attempt to explain morality and replace ethics with evolutionary theory.

I would venture to say that the received or official view certainly on the latter matter is a form of *scientific restrictionism*. Both scientists and philosophers have defended such a view. Albert Einstein, for instance, writes, 'As long as we remain within the realm of science proper, we can never meet with a sentence of the type "Thou shalt not lie" ... Scientific statements of facts and relations ... cannot produce ethical directives' (quoted in Singer 1981: 54). Likewise, Nicholas Rescher maintains,

> The characteristic cognitive task of science is the *description* and *explanation* of the phenomena—the answering of our *how?* and *why?* questions about the workings of the world. Normative questions of value, significance, legitimacy, and the like are simply 'beside the point' of this project. ... [To deny this and become an advocate of Scientism] is not to celebrate science but to distort it by casting the mantle of its authority over issues it was never meant to address.
>
> (Rescher 1984: 213)

Furthermore, Holmes Rolston claims,

> But science is never the end of the story, because science cannot teach humans what they most need to know: the meaning of life and how to value it. The sciences are as practical as theoretical; science has evident survival value, teaching us how to gain benefits that we desire. But what ought we to desire? Our enlightened self-interest? Our genetic self-interest? More children? More science? The conservation of biodiversity? Sustainable development? A sustainable biosphere? The love of neighbor? The love of God? Justice? Equity? Charity? ... After science, we still need help deciding what to value; what is right and wrong, good and evil; how to behave as we cope. The end of life still lies in its meaning, the domain of religion and ethics.
>
> (Rolston 1999: 161–2)

Do these recent developments in biological science give us reason to reconsider the merits of scientific restrictionism when it comes to issues of morality? In what way, more exactly, can morality be explained by evolutionary theory and is this the best explanation we can offer? Is it possible and desirable to develop a scientific evolutionary ethic? Or what, more exactly, is it that evolutionary biology offers when it comes to ethics? I shall claim that one can maintain that evolutionary theory can be of relevance for morality in different ways or, to phrase it differently, a 'biologicization' of ethics can mean a number of different things.[1] Although scientists like Wilson maintain clearly that evolutionary theory is of great significance to ethics, it is, nevertheless, less clear exactly what that significance consists of. Let me, therefore, start by suggesting some ways in which it is possible that evolutionary biology could be of importance for ethics:

(A) Evolutionary biology can *explain* the development and maintenance of morality in human life. It can give an account of why we behave morally and why we believe the ethical statements that we do.

(B) Evolutionary biology can provide us with new information about human life and its environment that can *undermine* (or support) existing ethical theories, norms or beliefs.

(C) Evolutionary biology can *justify* ethical norms or beliefs and provide us with a new ethic. It can tell us how we morally ought to behave, that is, what ethical norms we should follow and what kind of virtues we ought to develop.

Let us consider the merit of each one of these different ways in which evolutionary theory could be of relevance to morality or moral reasoning. In this chapter I shall focus on (A) and state the content of the evolutionary explanation of morality and some of its implications, followed by a critical assessment of it. In the next chapter the case for project (B) and (C) will be stated and also critically evaluated. In the process our understanding of the possibilities and limitations of the evolutionary explanation of morality will be sharpened.

The Darwinian account of morality

The task Wilson identifies for the new Darwinian social scientists is to explain the social behaviour of morality in evolutionary terms. The biologists know that human self-knowledge is constrained and shaped by the hypothalamus and limbic system of the brain and that these evolved by natural selection.

---

[1]   In this study I shall not (which is quite possible) discriminate between 'ethics' and 'morality' but use the two terms more or less interchangeably.

These insights, he writes, 'must be pursued to explain ethics and ethical philosophers, if not epistemology and epistemologists, at all depths' (Wilson 1975: 3). The key idea is that morality, like all social behaviour, is an outcome of an evolutionary process that selects some genes or groups of genes in preference to others. Richard D. Alexander maintains that the mistake done is that 'those who have tried to analyze morality have failed to treat the human traits that underlie moral behavior as outcomes of evolution – as outcomes of the process, dominated by natural selection, that forms the organizing principle of modern biology' (Alexander 1987: xiv). The question we have to address is how far this will take us in understanding and explaining ethics and moral behaviour.

Let us start by specifying the content of the Darwinian explanation of morality. According to Ruse and Wilson (1993: 308–9), it is based on two claims:

(1) 'The social behaviour of animals is firmly under the control of the genes, and has been shaped into forms that give reproductive advantages.'

and

(2) 'Humans are animals.'

The first claim, they tell us, has been repeatedly confirmed by 'a multitude of recent studies, from fruit flies to frogs' and evidence to confirm the second claim continues to pour in from almost every discipline within biology (Ruse and Wilson 1993: 309). We know, for instance, that at the biochemical level we are today closer relatives of the chimpanzees than the chimpanzees are of the gorillas.

Notice that these claims as they are stated, also commit Ruse and Wilson to a third claim, which is entailed logically by (1) and (2), namely:

(3) Human behaviour (including morality) is firmly under the control of the genes, and has been shaped into forms that give reproductive advantages.

Wilson more explicitly endorses (3) when he states elsewhere what the 'general sociobiological view of human nature' is. It is the view that 'the most diagnostic features of human behavior evolved by natural selection and are today constrained throughout the species by particular sets of genes' (Wilson 1978: 43). We may think that much human behaviour goes way beyond biology and genetics, but this is essentially wrong because behaviour is firmly under the control of the genes. Therefore, the 'question of interest is no longer whether human social behavior is genetically determined; it is to what extent. The accumulated evidence for a large hereditary component is more detailed and compelling than most persons, including even geneticists, realize. I will go further: it already is decisive' (19). Whether or not we realize it, we are, in Dawkins' words, 'survival machines – robot vehicles blindly programmed to preserve the selfish molecules known as genes' (Dawkins 1989: v).

But notice that even if we agree with these biologists that humans are animals, it is nevertheless true that we are also different from other animals in many ways. We are self-conscious, have developed the ability to think discursively and to express these thoughts in language, to do mathematics and science and so on. It therefore remains an open question *how much* of human social life can be explained in biological terms; how much of human behaviour is firmly under the control of the genes, and has been shaped into forms that give reproductive advantages. In particular, it still remains to be seen if evolutionary theory can tell us anything really interesting about morality (science, religion or any other distinctive human activity). Hence, the premises of Ruse's and Wilson's argument must be more carefully stated:

(1') The social behaviour of animals such as fruit flies, ants and frogs is firmly under the control of the genes, and has been shaped into forms that give reproductive advantages.

(2) Humans are animals.

(3') Therefore, it is likely that human behaviour (including morality) is firmly under the control of the genes, and has been shaped into forms that give reproductive advantages.

Let us also, for clarity, say that an 'animal' is any being who has evolved through natural history and belongs (or has belonged) to the tree of life, sharing common ancestry.

Whether there are good reasons to believe that (3') is true, is an issue we shall have to return to but it depends on how different humans are from other animals. My point here is merely that once we take seriously that human beings are part of the evolutionary scheme, we cannot deny the *legitimacy* of trying to explain why we have certain ethical ideas by relating them to our evolutionary history. So the search for a biological explanation of morality seems to fall within the scope of the natural sciences, and this explanation must be allowed to compete with any other explanation given of the phenomena of morality that can be found within the social sciences, the humanities or theology.

Thus, the central thesis of sociobiology or evolutionary expansionism can be stated as follows:

(T) The most dominant determinant in human behaviour is maximizing fitness, that is, the production of the most offspring in the following generations.

This is the meaning I shall assign to the claim that 'human behaviour is firmly under the control of the genes' (rather than the control of reason or culture). The explanatory scheme is:

(E) Trait *p*, behaviour *q* or institution *A* exists and continues to exist because it emerged and continues to function as a strategy (or part of a strategy) adapted to secure the fitness of the individuals and their genes.

But what is more exactly the content of the Darwinian explanation of the emergence of morality and of why we continue to be engaged in moral activities? Let us start with the theory of evolution. The key idea is that natural selection is the principal explanation of the evolution of the diversity of life forms we can find on earth. That is to say that those organisms with superior competitive abilities (for instance speed) are likely to survive longer and produce more offspring. This is true if (a) variation of abilities exist within the group of organism, (b) variation of abilities is variation in fitness (the capacity of survival and reproduction), and (c) the abilities can be inherited. Heritable variation in fitness produces evolution. The key unit of natural selection has often been thought to be the individual. Biologists like Wilson and Dawkins do not think that this is correct and propose instead that a trait (or an organism) evolved because it benefits the genes. Genes are instead the key unit of selection. They are taken to be the deeper causes of what happens in evolution. They build 'survival machines' (i.e. organisms) which evolved to promote the interests of the genes they contain (Dawkins 1989: v). Those genes that are best at replicating themselves in subsequent generations will be selected in the evolutionary process and the others will disappear.

Now scientific expansionists extend this whole idea from nature to culture, that is, natural selection is taken to be the principal explanation of human behaviours and traits. So the key to understanding and explaining moral behaviour is to ask in what way it has contributed and continues to contribute to reproduction, that is, to the survival of the genes. The claim is that morality exists and flourishes because it promotes the survival and multiplication of the genes that direct its assembly. As Ruse writes, 'Morality, or more particularly the moral sense, comes about because the moral human has more chance of surviving and reproducing than the immoral person' (Ruse 1985: 197). But how exactly did this come about? The biological hypothesis is that the human ability to be moral arose out of animal cooperation. Ruse and Wilson write that a 'number of causal mechanisms – already well confirmed in the animal world – can yield the kind of cooperation associated with moral behaviour' (Ruse and Wilson 1986: 178–9). The idea is that the evolution of morality can be explained by such mechanisms as kin selection and reciprocity.

We can find among animals behaviour that is performed for the benefit of others. This happens when, for example, a hawk flies overhead and blackbirds utter warning cries, thereby increasing the risk that they themselves will become the hawk's victims, and when other animals such as African wild dogs threaten or even attack predators to protect other members of their species. But if evolution is a struggle for survival, how can evolutionary theory explain these phenomena of self-sacrifice that seem to increase another animal's prospects of survival at the cost of their own? The answer is *kin selection*, that is, the genetically based tendency to help one's relatives. Self-sacrificing behaviour of this sort exists because, contrary to appearance, it benefits the real unit of selection, the genes. Those genes that are best at replicating themselves in subsequent generations will be selected in the evolutionary process and the others will disappear. Animals are after all organisms genetically programmed to preserve the genes. Hence, by sacrificing or risking to do so, the blackbird and the wild dog promote the spread of their own genes since their relatives share these with them. Risking their lives does not harm the prospects of their genes surviving if their behaviour eliminates a similar risk to the lives of two of their children, four of their nieces or eight of their first cousins. This kind of cooperation with kin is understood in terms of *inclusive fitness* in contrast to the *personal fitness* of the individual.

But not all animal behaviour is such that it is performed for the benefit of helping relatives. Monkeys groom each other even when they are not relatives. They and other animals such as wolves and chimpanzees often share food and cooperate in other ways to help each other, even if they are not related. Here evolutionary biologists use the idea of *reciprocity* as an explanation: if you scratch my back I'll scratch yours. Robert L. Trivers suggests that this kind of cooperative behaviour could develop in circumstances in which it was likely to aid both parties (Trivers 1971). Such cooperative behaviour is likely to leave more offspring than non-cooperative behaviour. It promotes personal and genetic fitness.

Scientific expansionists maintain now that the empirical evidence suggests that cooperation between human beings was brought about by the same evolutionary mechanism as animal cooperation (kin selection and reciprocity) and these determinants are in both animals and humans taken to be heritable. Thus, Wilson writes that in the course of evolutionary history 'genes predisposing people toward cooperative behavior would have come to predominate in the human population as a whole. Such a process repeated through thousands of generations inevitably gave rise to moral sentiments' (Wilson 1998: 59). People 'vividly experience these instincts variously as conscience, self-respect, remorse, empathy, shame, humility, and moral outrage' (59).

So by appealing to fitness-maximizing strategies biologists like Wilson maintain that they can explain behaviour like altruism, promiscuity, xenophobia, homosexuality, the honouring of one's parents, and institutions like the incest taboos, preferential marriage rules, sex role differences, social inequality and double standards. For instance, in culture we find moral prohibitions against incest. We find among people a moral belief saying that incest is morally wrong. These biologists claim that the maintenance of this moral belief can be explained by evolutionary theory as a result of fitness-maximizing strategies since incest results in lowered genetic fitness. The full explanation is that 'lowered genetic fitness due to inbreeding led to the evolution of the juvenile sensitive period by means of natural selection; the inhibition experienced at sexual maturity led to prohibitions and cautionary myths against incest or (in many societies) merely a shared feeling that the practice is inappropriate' (Ruse and Wilson 1986: 184). The idea is that over thousands of years humans who had genes that inclined them to avoid incest left more offspring in the following generations. Thus we have good reasons to think that the moral belief 'incest is morally wrong' is genetically fixed or favoured in human nature. Our genes predispose us to think that incest is morally wrong and to develop myths of both religious and philosophical kinds to support this belief. A support we thought provided the justification of the prohibition against incest, but which evolutionary theory shows is merely a kind of superstructure that our genes biased or fooled us into developing.

The claim is that the same explanatory pattern could be used to explain successfully all other forms of moral behaviour we can find in culture. Ruse and Wilson summarize their case by saying, 'In short, ethical premises are the peculiar products of genetic history, and they can be understood solely as mechanisms that are adaptive for the species that possess them' (Ruse and Wilson 1986: 186). Ruse and Wilson are thus ready to defend (T) and apply (E) to all *moral* reasoning, that is:

(T_m) The most dominant determinant in moral behaviour is maximizing fitness, that is, the production of the most offspring in the following generations.

(E_m) Moral trait p, moral behaviour q or moral institution A exists and continues to exist because it emerged and continues to function as a strategy (or part of a strategy) adapted to secure the fitness of the individuals and ultimately their genes.

(T_m) and (E_m) claim, of course, more than these biologists can deliver. This is something Ruse and Wilson are also willing to admit. What they and other evolutionary biologists attempt to do instead is to assemble more and more examples of moral beliefs and traits that can be successfully explained as fitness-maximizing strategies and in this way build an inductive case for the truth or probability of (T_m) and (E_m). This means, however, that (T_m) and (E_m) can be *disconfirmed* if we can find examples of moral beliefs and traits that cannot successfully be explained as fitness-maximizing strategies. Moral behaviour could exist that does not favour, and may even hinder, the survival and reproduction of the individuals and their genes.

A non-Darwinian account of morality

Let us now consider the merits of the Darwinian explanation of morality in comparison to other rival explanatory theories. I have argued that once we take seriously that human beings are a part of the evolutionary scheme, we cannot deny the legitimacy of trying to explain why we have certain ethical ideas or behave morally by relating them to our evolutionary history. So the search for a biological explanation of morality falls within the scope of the natural sciences, and this explanation must be allowed to compete with any other explanation given of the phenomena of morality that can be found within the social sciences, the humanities and theology.

It is nevertheless true that philosophers typically have not paid much attention, at least until recently, to the issue of the biological origin of morality and in that sense have tacitly presupposed that such an explanation is not really all that relevant to what they consider to be most important, namely the justification of moral theories, norms and beliefs. To understand why this has been so and whether or not it is justified, it is important to take into account what moral philosophers consider ethics and moral behaviour to be. Thomas Nagel expresses what I think is a typical philosophical understanding of ethics when he writes,

> The usefulness of a biological approach to ethics depends on what ethics is. If it is just a certain type of behavioral pattern or habit, accompanied by some emotional responses, then biological theories can be expected to teach us a great deal about it. But if it is a theoretical inquiry that can be approached by rational methods, and that has internal standards of justification and criticism, the attempt to understand it from outside by means of biology will be much less valuable.
>
> (Nagel 1979: 142)

So while it is true that ethics begins with behavioural patterns or habits or pre-reflective ideas about what to do, how to live and how to treat other people, philosophers typically believe that ethics nevertheless subjects these patterns, habits or ideas to criticism and revision and creates new ideas and forms of conduct. But, as Nagel is aware, this view of morality presupposes that ethical beliefs and patterns of behaviour

> can be criticized, justified, and improved – in other words that there is such a thing as practical reason. This means that we can reason not only, as Hume thought, about the most effective methods of achieving what we want, but also about what we should want, both for ourselves and for others.
>
> (Nagel 1979: 144–5)

It is, of course, on this point that the biologists we have considered think evolutionary theory can be of great relevance because if philosophers had paid attention to biology they would have realized that this view of ethics is scientifically uninformed and therefore ought to be rejected. There is no such thing as practical reason in this sense. Whether or not we are aware of it, our moral reasoning is done with the aim of maximizing fitness. This is so since these biologists believe, as we have seen, that evolutionary theory supports the thesis:

(T)   The most dominant determinant in human behaviour is maximizing fitness, that is, the production of the most offspring in the following generations

and (T) entails,

(T_m)   The most dominant determinant in *moral* behaviour is maximizing fitness.

Thus there is no room for practical reason.

However, the sociobiological research programme on this point undermines itself because it leaves no room for theoretical reason either. Since (T) entails (T_m), it also entails,

(T_s)   The most dominant determinant in *scientific* behaviour is maximizing fitness.

Hence it follows that the dominant determinant of why biologists like Alexander, Dawkins, Ruse and Wilson launched this research programme is that they are trying to maximize fitness, foremost their own individual fitness, secondly, their genetic fitness and when necessary secure these forms of fitness through reciprocal behaviour. Unreflectively we perhaps thought that they advocated this explanatory expansion of evolutionary theory into the social sciences, the humanities and theology primarily because they believe that the scientific evidence supports this view. This, however, cannot be the best explanation of their behaviour according to their own theory because such an explanation is not supported by (T_s). But then, of course, we have no reason to believe that their claims about morality are true because what they really are up to is maximizing fitness. Hence, their own attempt to scientifically justify sociobiology presupposes that (T_s) is false, both in the sense that they can give rational (or truth-apt) arguments for why the scientific community ought to accept their theory as a part of the body of scientific knowledge, and in the sense that the members of the community are able to understand and evaluate these arguments as rational arguments.

Therefore, if these biologists want to avoid the embarrassing consequence that their theory falsifies itself, they have to admit that in human behaviour there are important *exceptions* to the Darwinian explanation scheme. Or, rather, that is one way out of this dilemma. Another way is to weaken the general claim by giving up the idea that human behaviour is firmly under the control of the genes, that is, that we are survival machines programmed to preserve 'selfish' genes. Instead our biological nature merely predisposes us to behave and reason in a certain way, but it does not control our moral and intellectual behaviour. Our genes influence our behaviour, but influence and predisposition is not the same as determination and control. This would make it possible for people to behave morally altruistically when such action comes in conflict with biological selfishness. In a similar way, even though it has perhaps been difficult for Alexander, Dawkins, Ruse and Wilson to put aside as irrelevant the issue of whether this expansion of evolutionary theory increases their own fitness and focus on the evidence, they have in this case done so successfully; and, therefore, their account of moral behaviour is correct.

So they can avoid inconsistency by modifying (T), and instead maintain

(T′) Human behaviour is predisposed (but not controlled or determined) by our genes to maximizing fitness, that is, the production of the most offspring in the following generations.

Or they can stick to the idea that our behaviour is in general firmly under the control of the genes and determinate to maximize fitness, but because of some peculiar reason or perhaps by pure blind chance, this is not true about the behaviour of natural scientists. Hence,

(T″)The most dominant determinant in human behaviour, with the exception of the natural sciences, is maximizing fitness, that is, the production of the most offspring in the following generations.

According to (T″), people (or some people at least) are capable of entertaining and evaluating scientific theories *independently* of whether these beliefs and practice optimize their numbers of offspring in the next generation. But (T″) also implies that this is not the case when it comes to ethical or legal principles, religious or political convictions and so on. Practical reason but not theoretical reason is the slave of the genes.

But critics can argue that the best explanation of much moral behaviour is the very same as the explanation of much scientific behaviour. Just as the intellect can do more than support survival and reproduction in science, so it can do this in ethics or any other human affairs. The explanation is that humans have developed over evolutionary time eminent cognitive abilities, an intellect, which includes self-awareness and abstract and critical thinking. It is the possession of this ability or faculty that allows the development of science, ethics, religion or any other distinctive human activity. With the evolution of human beings, natural history has reached a point at which certain creatures evolved who could question their standing in the world in a way that no other part of nature can. It is our self-consciousness that makes this possible. In the interaction with other people and nature, we become aware that we are knowers and agents, and thus begin to understand that our beliefs and practice are ours. We realize that our beliefs and practices are, therefore, not necessarily in tune with those of others or with the world, that our path through the world is but one possible path. Thus our self-consciousness initiates a process which leads us to attempt to revise and improve our beliefs and to ask questions about which courses of action are for the best.[2]

It is thus our *intelligence* that provides us with the means, in science, to evaluate theories and, in morality, to evaluate certain actions as either good or evil, and to choose to believe and do other things than those, which maximize our offspring in the next generation. Peter Singer develops this alternative explanation in the following way:

> Gradually, as we evolved from our pre-human ancestors, our brain grew and we began to reason to a degree no other animals had achieved. We became better able to communicate with our fellows. Our language developed to the point at which it enabled us to refer to indefinitely many events, past, present, or future. We became more aware of ourselves as beings existing over time, with a past and a future, and more conscious of the patterns of our social life. We could reflect, and we could choose on the basis of our reflections. All this gave us, of course, tremendous advantages in the evolutionary competition for survival.
>
> (Singer 1981: 91)

The biologist Francisco Ayala agrees with Singer, arguing that increased intelligence made it possible for our human ancestors (a) to anticipate the consequences of their own actions, (b) to make value judgments; and then (c) to choose between alternative courses of actions (Ayala 1987: 237). He furthermore maintains that these factors that are required for intelligence in general are also required for consciousness and ethical behaviour. He writes,

> Only if I can anticipate that pulling the trigger will shoot the bullet, which in turn will strike and kill my enemy, can the action of pulling the trigger be evaluated as nefarious. ... Only if I can see the death of my enemy as preferable to his survival (or vice versa) can the action leading to his demise be thought as moral. ... Pulling the trigger can be a moral action only if I have the option not to pull it. ... [T]he circulation of the blood or the process of food digestion are [thus] not moral actions.
>
> (Ayala 1987: 237–9)

The critics of sociobiology seem to agree up to this point. But they diverge on the issue of whether not only intelligence but also ethical behaviour (or consciousness) is an attribute directly promoted by natural selection. Ayala does not think that ethical behaviour developed because it was adaptive in itself. He tells us that he finds 'it hard to see how *evaluating* certain actions as either good or evil (not just choosing some actions rather than others, or evaluating them with respect to their practical consequences) will promote the reproductive fitness of the evaluators' (Ayala 1987: 239). His theses are therefore, that 'ethical behavior is not causally related to the social behavior

---

2   An interesting development and justification of these ideas can be found in O'Hear (1997: 16–46).

of animals, including kin and reciprocal "altruism"' and that 'moral norms ...
are products of cultural evaluation, not of biological evolution' (236–7).
Others like Singer and Rolston think it is quite plausible that ethics is adaptive
(see Rolston 1999: 263). Singer, in discussing the chances egoists and altruists
would have of surviving if attacked by a sabertooth cat, writes,

> Evolution would ... favor those who are genuinely altruistic to other
> genuine altruists, but are not altruistic to those who seek to take advantage
> of their altruism. We can add, again, that the same goal could be achieved
> if, instead of being altruistic, early humans were moved by something like
> a sense that it is wrong to desert a partner in the face of danger.
>
> (Singer 1981: 49)

Which one of these non-Darwinian theories of morality is the more plausible
is not necessary to determine in this context. What is important is, as Philip
Kitcher points out, that it is 'possible to take the evolution of *Homo sapiens*
seriously and yet to deny that natural selection has fashioned dispositions to
behavior that lead us always (or almost always) to maximize our inclusive
fitness' (Kitcher 1986: 402). Hence, I take a theory to be 'non-Darwinian' if
it denies that the most dominant determinant in human behaviour and
reasoning is maximizing fitness.

But even if it is possible to take the evolution of *Homo sapiens* seriously
and yet to deny that natural selection has fashioned dispositions to behaviour
that lead us always (or almost always) to maximize our inclusive fitness, is
there any reason why we should prefer a non-Darwinian account of morality
to the rival Darwinian (or sociobiological) account? Why believe that if the
intellect can do more than support survival and reproduction in science, it can
do this also in ethics? We cannot on this issue maintain that Wilson, Ruse and
others have developed a theory that is self-referentially incoherent. The
approach must instead be to provide evidence that undermines the plausibility
of (T''), namely that the most dominant determinant in human behaviour,
with the exception of the natural sciences, is maximizing fitness.

Can we find any convincing examples of non-scientific human behaviour
that cannot successfully be explained as fitness-maximizing strategies? The
most obvious is that while it is true of most species that every organism has
as many descendents as it can, it is not true of ours. In modern Western
societies parents have fewer children than they could successfully raise given
their economy. But then the most dominant determinant in human
reproductive behaviour cannot be maximizing fitness, and the explanation for
the fall in fertility in the West cannot be genetic but cultural. The most
reasonable explanation is that we, as a result of scientific discoveries, can by
using contraception choose the number of children we want to have. So it
cannot be true that in our sexual behaviour we are determined by our genes to
increase our offspring, nor does it seem that – if we weaken the claim to (T')

– that our predisposition in this case is very strong. If we are strongly predisposed by our genes, it is in respect of our desire for sexual pleasure. But because of our eminent intellectual resources we are beings capable of anticipating the consequences of our actions and of developing means to obtain sexual pleasure without reproduction, and so many people in the West have chosen to do. These facts cause profound problems for an extension of evolutionary theory to human behaviour because maximizing fitness is the very 'motor' of evolution. But in this crucial case it does not 'run' as the extended evolutionary theory predicts it would. Voluntary restrained reproduction constitutes powerful counter-evidence to (T) or (T″).

These biological expansionists could, of course, reply that this is merely an anomaly in natural history because such behaviour is self-eliminating. Those who maximize their fitness will in subsequent generations replace people who limit their offspring. But as Holmes Rolston points out, this does not follow,

> because the idea that one ought to have fewer children is not itself genetically transmitted, as is proved by the fact that the couples now restraining reproduction are only a generation or two removed from parents and grandparents who had other ideas. If the idea is contagious enough culturally, and if it is appealing for good reasons, it can spread indefinitely through the population, jumping genetic lines and at a speed of transmission many orders of magnitude faster than any behavioral tendencies transmitted genetically. Those with the new cognitive beliefs convert the oncoming generation to their view.
>
> (Rolston 1999: 132)

Now we are beginning to see that our eminent cognitive resources make it possible for us to spread ideas, scientific theories, moral and religious convictions and the like *non*-genetically, through *cultural* transmission or communication. People teach each other how to do and how to think about things such as evolutionary theory, the golden rule, growing wheat and baking bread. Such ideas are discovered in the past and transmitted non-genetically from parents to children, from teachers to students and so on. But as anthropology, religious studies and so forth have shown, parents and teachers teach children and students different things in different cultures; and we can by travelling or reading books about these things learn what cultures other than our own believe about abortion, human rights, God or love. Hence, many of the things we do or believe seem to depend on education and cultural upbringing rather than our genes.

Rolston suggests that a model we can use to understand the relationship between biology and culture is the hardware-software model:

> The human cognitive equipment has what structure it has, like a computer hardware, as a given to work with. Quite diverse software programs can

> be run on this hardware, and, in terms of selecting among the broad
> cultural options faced, nothing is hard-wired. ... When humans choose
> between competing options, those who use this and not that software will
> better succeed in reproducing, and their children will inherit copies of it.
> This better capacity to survive is copied, but not by hardware rebuilding
> (not by genetics), rather software duplication (cultural transmission).
> There is no reason to call this Darwinian selection. This is Lamarckian
> selection, since acquired traits are being transmitted. Better still, let us
> simply call it cultural selection and recognize that it is paralleling and
> even transcending biology.
>
> (Rolston 1999: 139)

It is, of course, important how the hardware evolved and it sets some
limitations for what software can be used, but as Rolston insists,

> the cultural software that is run on the biological hardware, does make a
> critical difference. Consider, in the Western heritage, the rise of Israel, the
> crucifixion of Jesus, the signing of the Magna Carta, the rise of science,
> Martin Luther and the Protestant Reformation, the defeat of the Spanish
> Armada, the American Revolution, James Watt's inventing the steam
> engine and revolutionizing transportation, Abraham Lincoln's setting the
> slaves free, the discovery of nuclear weapons, or the computer revolution.
> Without genes, none of these events would have taken place. But with our
> genes none is explained in contrast to other events that might have taken
> place but did not – or in contrast to the different events that took place in
> other cultures elsewhere on Earth.
>
> (Rolston 1999: 155)

Rolston is, of course, not alone in suggesting events that appear to be
unexplainable by biological categories. David Stove maintains that we can
easily think of a hundred characteristics of human behaviour that are of this
sort. Stove starts off with the letter 'a': abortion, adoption, fondness for
alcohol, altruism, anal intercourse, respect for ancestors, the importance
attached to art, asceticism, whether sexual, dietary or whatever (Stove 1994:
275). Anthony O'Hear lists virtues such as feeding the poor, tending the sick,
visiting the imprisoned, modesty, chastity, honesty, promise-keeping,
integrity, respect for the rights of others, self-sacrifice, honour, faith, hope and
charity, and concludes that 'those virtues may presuppose another non-
Darwinian world for their intelligibility' (O'Hear 1997: 143). These authors
all agree that non-scientific behaviour in general and moral behaviour in
particular exists that does not favour, and sometimes may even hinder, the
survival and reproduction of individuals and their genes.

We cannot, of course, examine all of these cultural phenomena to
determine whether or not they can plausibly be explained by evolutionary
theory alone. What we can do is simply to look a little more closely at some
of our moral convictions. For instance, many parents have a strong moral
conviction that they are responsible for their children. These parents think

they morally fail if they do not give their children love and support and nurture them until they can take care of themselves. But, as Dawkins points out, from the point of view of the 'selfish' genes there is no distinction of relevance between caring for a baby brother and caring for a baby son because both infants are equally closely genetically related to the parents (Dawkins 1989: 110). Therefore, it would be evolutionarily advantageous for mothers to seek to have their children adopted by other women so that they themselves can have more children and maximize their genetic fitness. Hence if we truly are survival machines for our genes, if our moral behaviour is firmly under the control of the genes, then mothers would whenever possible have their children adopted. But this is clearly not the case. Therefore, it is false that, with the exception of the natural sciences, the most dominant determinant in human behaviour is maximizing fitness.

Another related example is that humans do not have a moral conviction saying 'Sterile and homosexual humans ought to devote their lives to the offspring of their brothers and sisters.' But if ethics is, as these biologists will have us to believe, an illusion fobbed off on us by our genes, then surely our genes would have tricked us into thinking that sterile and homosexual humans have a very special responsibility for the offspring of their brothers and sisters because such behaviour would maximize genetic fitness. Nor can we find among people a moral belief saying 'One ought to be ready to sacrifice one's life for more than two brothers, or four half-brothers, or eight first cousins.' But if we truly are survival machines for our genes, then we can expect that we are programmed in such a way that those that cannot, or for other reasons will not have any children of their own, would be tricked by their genes to promote the survival of their close genetic relatives. In fact, this is what one of the founders of the sociobiological research programme admits. W. D. Hamilton writes that 'no one is prepared to sacrifice his life for any single person, but ... everyone will sacrifice it [his life] for more than two brothers, or four half-brothers, or eight first-cousins' (quoted in Stove 1994: 269). But this is just not the case. If extending evolutionary theory to moral behaviour is right, we can predict that certain types of behaviour should be found among people. But we cannot find these types of behaviour (at least they are not as widely spread as we would expect if the theory were true). There exist, therefore, good reasons to believe that the extended evolutionary theory (or more exactly T″) is incorrect or incomplete and an additional non-Darwinian explanation of moral convictions is needed.

Lastly, let us focus on some of the moral issues in contemporary ethics. The examples discussed will be taken from the field of environmental ethics. A number of philosophers, theologians and scientists have recently argued that we ought to reject our traditional anthropocentric environmental ethics and instead adopt non-anthropocentrism. We should stop believing that people's behaviour towards nature should ultimately be evaluated on the basis of how it affects human beings and only such beings, and instead adopt the moral conviction that people's behaviour towards nature should be evaluated also on the basis of how it affects *other* living beings. This ought to be the case because they think that not only humans, but also other things in nature have moral standing and can consequently be treated morally rightly or wrongly. A number of these non-anthropocentrists also argue that we have a duty towards our fellow beings to limit significantly the number of humans living on this planet. Rolston maintains that 'conserving the Earth is more important than having *more* people'. It is even 'more important than the needs, or even the welfare, of *existing* people' (Rolston 1994: 233). Furthermore, Arne Naess writes as a response to an answer given by a United Nations study (to the question 'Given the present world-wide industrial and agricultural capacity, technological development, and resource exploitation, how many people could be supported on earth today with the standard of living of the average American?') that,

> The authors think that 500 million would not result in a uniform, stagnant world and refer to the seventeenth century. Agreed, but the question raised refers only to humans. *How about other living beings*? If their life quality is not to be lowered through human dominance, for instance agriculture, are not 500 million too many? Or: are cultural diversity, development of the sciences and arts, and of course basic needs of humans not served by, let us say, 100 million?
>
> (Naess 1989: 140–1, emphasis added)

Lastly, Irvine and Ponton list a number of options available to implement this kind of population limitation policy. They write,

> There could be payments for periods of non-pregnancy and non-birth (a kind of no claims bonus); tax benefits for families with fewer than two children; sterilization bonuses; withdrawal of maternity and similar benefits after a second child; larger pensions for people with fewer than two children; free, easily available family planning; more funds for research into means of contraception, especially for men; an end to fertility research and treatment; a more realistic approach to abortion; the banning of surrogate motherhood and similar practices; and the promotion of equal opportunities for women in all areas of life.
>
> (Irvine and Ponton 1988: 23)

Can evolutionary theory, as understood by scientific expansionists, explain why these people have these moral convictions? Can it explain why non-anthropocentrists believe that (a) other beings besides humans have moral standing, that (b) we have a moral duty to ensure that the size of the population is reduced to a level that is compatible with a respect for other living things and the integrity (or dignity) of species and ecosystems, and that (c) a good way of obtaining this objective might be to offer payments for periods of non-pregnancy and non-birth, tax benefits for families with fewer than two children, sterilization bonuses and so on? How could one believe that 'the overarching goal of wildlife management should not be to insure maximum sustainable yield; it should be to protect wild animals from those who would violate their rights' (Regan 1983: 357) or that 'In every case the effect upon ecological systems is the decisive factor in the determination of the ethical quality of actions' (Callicott 1989: 21), if the most dominant determinant in morality is the production of most offspring in the next generations? It seems as if the 'selfish' genes have lost their grip over these people since contrary to the 'interests' of their genes they are ready to limit their offspring significantly for the sake of living things other than humans.

Thus, there exist good reasons to believe that our eminent cognitive resources can do more than support survival and reproduction, not only in science, but also in morality. In fact, moral convictions exist that do not favour but even hinder the survival and reproduction of individuals and their genes. People do have norms and moral practices that are significant in their lives and thought and either indifferent to survival in their effects or actually counter to survival. Hence, it seems that a non-Darwinian explanation is more plausible and compatible with the evidence than a Darwinian explanation.

The focus of this chapter has been on the attempt by certain evolutionary biologists to explain the development and maintenance of morality in human life. There is no doubt that this project falls within the scope of science. Once we accept that humans are also animals in the sense that they have a common ancestry with all other living things on this planet, then it must be appropriate to investigate to what extent explanations of the behaviour of animals can be applied to humans as well. But we have seen that the Darwinian explanation, that morality exists and continues to exist because it emerged and continues to function as a strategy (or part of a strategy) adapted to secure the fitness of the individuals and their genes, has limited explanatory scope. Vast stretches of the moral landscape cannot plausibly be explained as fitness-maximizing strategies. In fact, moral convictions exist, for instance in environmental ethics, that do not favour but even hinder the survival and reproduction of individuals and their genes.

A better explanation of much moral belief and behaviour is therefore that gradually, as we evolved from our pre-human ancestors, our brain grew and we began to reason to a degree no other animals had achieved, and it is the possession of this ability that make possible not only science, but also morality and our questions about which courses of action are for the best. These ideas about which courses of action are for the best are then spread non-genetically through cultural transmission or communication, and people have become convinced that slavery is morally wrong, that women ought to be given equal opportunities and responsibilities to men, that other living things besides humans have moral standing and so forth.

And not least importantly, people are able to evaluate these moral convictions and act upon them independently of whether they maximize fitness. Consequently, the account that these biological expansionists give of our moral capacities involves a radical and unsustainable redescription of what we are and what we do when engaged in moral reasoning. Anthony O'Hear's claim, 'for even though we and our capacities may have evolved in Darwinian ways, once evolved we and our capacities take off in quite unDarwinian ways', is thus at least right in respect to science and morality (O'Hear 1997: 214). Whether this is also true about religion will be the topic of Chapters 5 and 6.

CHAPTER FOUR

Debunking and replacing traditional ethics with science

In this chapter we shall take a closer look at two ways other than providing an explanation of ethics and moral behaviour (that is, project A) in which it is possible for evolutionary theory to be of great relevance for ethics. This will provide further insights into the strengths and weaknesses of a Darwinian explanation of ethics. A second way in which science in general or evolutionary biology in particular could be of significance for ethics is that it can provide us with new information about human life and its environment which either directly or in conjunction with extra-scientific claims undermines existing ethical theories, norms or beliefs (that is, project B). A third way (project C), which is more radical and challenging, is that science, that is, evolutionary biology, could tell us how to behave morally; we can replace traditional ethics with scientific ethics. Let us start by considering the merits of (B) and then move on to (C).

Debunking traditional ethics

Project (B) envisions perhaps the most obvious way in which science can be of relevance for ethics. But it is also the least controversial because it is in general accepted by philosophers and theologians who are involved in moral reasoning and justification. To see this consider Wilson's argument that human sexuality can be much better understood if we pay attention to new advances in evolutionary theory. Each society must make a series of choices on these issues concerning, for instance, sexual discrimination and acceptable sexual behaviour. These new advances in evolutionary theory 'are enough to establish [for instance] that the traditional Judeo-Christian view of homosexual behavior is inadequate and probably wrong' (Wilson 1978: 146). The argument is, roughly, that homosexual behaviour is found in nature, it is even likely that it has a genetic component, and thus is as fully natural as heterosexual behaviour. Hence, opposition on the grounds that homosexuality is unnatural cannot be defended.

Although this kind of biological information can be very important, it is nevertheless true that almost any scientific discipline can be of relevance for ethics in this sense. Ethicists are also aware that well-informed moral judgments must not be based merely on appropriate ethical norms but also on relevant and reliable information. So if this is all Wilson's talk about a biologization of ethics amounts to, then there is no justification for the grandiose claim that the humanities, including moral philosophy, are the last branches of biology to be included in the Modern Synthesis (Wilson 1978: 90). Furthermore, scientific information of this sort does not directly entail that value judgments or ethical norms are false. Even if Wilson's kin-selection hypothesis of the origin of homosexuality is correct, people may still believe that homosexuality is wrong, but they will have to find some other reason for holding this belief than that it is unnatural. Thus, the most Wilson is able to do in cases like this is to use evolutionary theory to undermine the original justification of the value judgment and not the value judgment itself. Science can merely in an indirect way undermine ethical convictions. This is something that has already been pointed out by other philosophers such as Peter Singer (1981) and Philip Kitcher (1986), but what is interesting, and what I want to focus my attention on, is that some of the claims we find in the writing of evolutionary biologists are of a more profound nature because they seem to directly entail that certain ethical theories, norms and beliefs are wrong. These are claims about the nature of morality and of the status of moral claims, namely that (1) *morality is ultimately a matter of selfishness* and that (2) *the objectivity of morality is an illusion*.

Of profound importance is that the advocates of the Darwinian explanation of morality take the explanation to undermine the conception of morality employed by philosophers and theologians. Philosophers and theologians (and I think also people in general) typically contrast morality with selfishness because morality is understood to require taking account of the interest of others, whether or not such behaviour serves one's own interest. Tom L. Beauchamp, for instance, writes, 'Normative ethics presupposes that one *ought* to behave in accordance with certain moral principles, whether or not such behavior promotes one's own interest' (Beauchamp 1982: 57). Anthony O'Hear maintains that 'morality, if it is to be morality, has to rest on principles of greater generality than what merely happens to conduce to my (or our) well being at a given time' (O'Hear 1997: 139).

But, as Ruse and Wilson maintain, if 'morality ... is merely an adaptation put in place to further our reproductive ends' or if 'in an important sense, ethics ... is an illusion fobbed off on us by our genes to get us to cooperate', then morality ultimately seems to be about self-interest or enlarged self-interest (that is, inclusive fitness) (Ruse and Wilson 1993: 310, cf. 1986: 186). Alexander is also very explicit about this consequence. He thinks that 'the true realization of intellectual advances in biological theory comes from their eventual application to human conduct: from their effect on humanity's view of itself' (Alexander 1987: 3). This change in 'self-view' includes 'the realization that ethics, morality, human conduct, and the human psyche are to be understood only if societies are seen as collections of individuals seeking their own self-interests (albeit through use of the group or group cooperativeness ...)' (Alexander 1987: 3). Alexander continues,

> these ideas run contrary to what people have believed and been taught about morality and human values: I suspect that nearly all humans believe it is a normal part of the functioning of every human individual now and then to assist someone else in the realization of that person's own interests to the actual net expense of those of the altruist. What this 'greatest intellectual revolution of the century' tells us is that, despite our intuitions, there is *not a shred of evidence to support this view of beneficence*, and a great deal of convincing theory suggests that any such view will eventually be judged false. This implies that we *will have to start all over again* to describe and understand ourselves, in terms alien to our intuitions, and in one way or another different from every discussion of this topic across the whole of human history.
>
> (Alexander 1987: 3, emphasis added)

Furthermore, these evolutionary biologists also have the answer to why our intuitions should have misinformed philosophers and theologians about the true character of morality. Ruse and Wilson tell us that the explanation is that

> human beings function better if they are deceived by their genes into thinking that there is a disinterested objective morality binding upon them, which all should obey. We help others because it is 'right' to help them and because we know that they are inwardly compelled to reciprocate in equal measure. What Darwinian evolution theory shows is that this sense of 'right' and the corresponding sense of 'wrong,' feelings we take to be above individual desire and in some fashion outside biology, are in fact brought about by ultimate biological processes.
>
> (Ruse and Wilson 1986: 179).

Natural selection has tricked humans to cooperate with others and care about them by prompting us into thinking that such behaviour is morally right. Therefore, 'ethics is a *shared* illusion of the human race. If it were not so, it would not work' (Ruse and Wilson 1993: 310). Moral norms and beliefs are legitimated by the illusion of objectivity.

Hence, even if project (C), the idea that evolutionary theory itself can generate and justify ethical norms, were to fail, the new information it gives (that is, project B) can, if true, undermine two fundamental presuppositions of much philosophy and theology, namely that morality is not about self-interest or inclusive fitness and that morality can be given a rational basis and thus in an important sense can be objective. If this new information proves to be correct then the application of evolutionary theory to morality has revolutionary consequences for ethics and a profound reconception of moral reasoning is required. Let us start by considering the merits of the first claim and then proceed to the second one in the following section.

The selfish gene theory

To be able to support the idea that morality and moral behaviour is ultimately about self-interest or enlarged self-interest these evolutionary biologists must show, or so it seems, that altruistic behaviour is also behaviour that serves such ends. If they fail to do this then the ordinary conception of morality can still be valid, that is, that we ought to behave in accordance with certain moral principles and develop certain moral virtues whether or not such behaviour promotes our own interests. If they succeed, we must stop thinking that we can behave altruistically, and develop – as Alexander states above – the 'self-view' that we and other people are individuals seeking their own self-interests and ultimately nothing else. We must start all over again to describe and understand our moral behaviour and ourselves.

So what about altruism? Alexander's general claim is, as we have seen, that 'there is not a shred of evidence to support' the view that it is 'a normal part of the functioning of every human individual now and then to assist someone else in the realization of that person's own interests to the actual net expense of those of the altruist' (Alexander 1987: 3). But not merely is it false that it is a normal part of our functioning, but the very paradigm examples of altruism we cite, such as the behaviour of people like Mother Teresa and anonymous blood donors, can be shown to be merely examples of kin selection or reciprocity.

The philosopher Peter Singer has tried to demonstrate the limits of a Darwinian account of morality by arguing that people who give blood anonymously do it to help others, and not out of any disguised desire to benefit themselves (Singer 1981: 133). Hence, the behaviour of anonymous blood donors is altruistic. Alexander's response is that giving blood may yield benefits from society for those who are known to be blood donors. 'For this very reason, a known blood donor may receive his "payment" from the members of society who accept him in social interactions or treat him deferentially compared to others known not to be blood donors or not known to be blood donors' (Alexander 1987: 157). In short, blood donation can be explained in terms of 'indirect reciprocity' that is, rewards from society at large. Furthermore, Alexander maintains that philosophers like Singer should not believe that 'common sense' is the right basis for understanding this kind of behaviour because it 'does not take into account well established biological facts and theories' (159). Thus, Alexander's conclusion is that blood donation is not altruistic behaviour.

Likewise, Wilson suggests that Mother Teresa's work of caring for the poor and ill of Calcutta may have appeared to be altruistic, but it was in fact not. Wilson writes about Mother Teresa's behaviour that,

> in sobering reflection, let us recall the words of Mark's Jesus: 'Go forth

to every part of the world, and proclaim the Good News to the whole creation. Those who believe it and receive baptism will find salvation; those who do not believe will be condemned.' There lies the fountainhead of religious altruism.'

(Wilson 1978: 165)

In short, Mother Teresa did these things because, as a Christian, she expected to be rewarded in heaven. Selfishness is the ultimate explanation to be offered for her behaviour. Wilson also raises the more principal question of whether culture can alter human behaviour from selfishness to altruism and he tells us that the 'answer is no' (165).

At this point it seems as if we find a direct conflict between the conception of morality found among philosophers and theologians on the one hand, and evolutionary biologists like Alexander, Ruse and Wilson on the other. What, however, makes it hard to know if there is a genuine conflict is that these biologists define 'selfishness' and 'altruism' quite differently from how philosophers and theologians define them. So to be able to compare these different accounts of morality with each other, we first have to be clear about how we ought to understand these concepts. Let us do this by examining the so-called 'selfish gene theory'.

Sociobiologists assume that the theory that (1) genes are the unit of natural selection in evolution implies that (2) genes and everything they have programmed is selfish. These claims are usually not distinguished, but let us, for reasons that soon will become evident, call the first the *gene selection theory* and the second the *selfish gene theory*. Dawkins expresses the content of the selfish gene theory as follows:

> Humans and baboons have evolved by natural selection. If you look at the way natural selection works, it seems to follow that anything that has evolved by natural selection should be selfish. Therefore we must expect that when we go and look at the behaviour of baboons, humans, and all other living creatures, we will find it to be selfish.
>
> (Dawkins 1989: 4)

He thinks that this selfish gene theory is a true but nevertheless 'astonishing' claim (Dawkins 1989: v). Other biologists make claims similar to those of Dawkins. George C. Williams, for instance, maintains, 'Evolution is guided by a force that maximizes genetic selfishness' (Williams 1988: 391).

But what exactly does it mean to be selfish, and what is the alternative? Dawkins tells us that 'an entity, such as a baboon, is said to be altruistic if it behaves in such a way as to increase another such entity's welfare at the expense of its own. Selfish behaviour has exactly the opposite effect. "Welfare" is defined as "chances of survival"' (Dawkins 1989: 4). Wilson says, 'When a person (or animal) increases the fitness of another at the expense of his own fitness, he can be said to have performed an act of *altruism.* ... In contrast, a person who raises his own fitness by lowering that of others is engaged in *selfishness*' (Wilson 1975: 117).

This would give us the following definitions:

> An entity is *biologically selfish* if it behaves in such a way as to decrease another entity's chances to survive and reproduce to the advantage of its own reproductive fitness.

> An entity is *biologically altruistic* if it behaves in such a way as to increase another entity's chances to survive and reproduce to the disadvantage of its own reproductive fitness.

Dawkins is aware that selfishness in this biological sense must be distinguished from its moral sense in that it does not presuppose any concept of intention or motive, but focuses solely on the effect of the behaviour. This is necessary because if selfishness presupposes motive or intention, and thus consciousness, then only human beings and not genes are able to behave selfishly or altruistically. Notice also that selfishness or egoism in the biological sense differs in a second way from its moral sense. An entity can only be morally selfish if it can choose between different options. A person, for instance, cannot be blameworthy of something he or she cannot in any way avoid doing. 'Ought' implies 'can'. But, of course, genes do not have any options in this sense, so this feature of moral selfishness must not be read into evolutionary theory.

We could, to make the differences clear, define 'selfish' and 'altruistic' in the moral sense as follows:

> People are *morally selfish* if they are motivated to satisfy their own self-interest without regard for others.

> People are *morally altruistic* (or *unselfish*) if they are motivated to benefit other even at their own expense or inconvenience.

This distinction attempts to discriminate, as Robert B. Ashmore points out,

> between those who devote their lives to caring for lepers and those whose career is loan-sharking. The ordinary person perceives a difference between sharing books with others and pilfering books from the public library. Few people confuse the motive of a parent assiduously attending to a sick child with the motive of one who beats a child that disturbs his sleep. Our customary use of these terms *selfish* and *unselfish* is intended to mark the real difference between these sorts of actions, motives, and states of character.
>
> (Ashmore 1987: 51)

Ashmore goes on to say that if we abandon this distinction and reduce everything to selfishness, it would require that we invent a new distinction to reflect everyday experiences of these differences in motive.

We have, then, two different sets of definitions, but is there a problem with this? The definitions that Dawkins and Wilson offer are probably acceptable stipulations of a biological use of the concepts 'selfishness' and 'altruism'. The serious problems start, however, when they and other evolutionary biologists say things such as:

> Much as we might wish to believe otherwise, universal love and the welfare of the species as a whole are concepts that simply do not make evolutionary sense. ... Be warned that if you wish, as I do, to build a society in which individuals cooperate generously and unselfishly towards a common good, you can expect little help from our biological nature.
>
> (Dawkins 1989: 2–3)

In what sense exactly are the words 'unselfishly', 'cooperate generously', 'universal love' and 'welfare' used in these sentences? What Dawkins writes does not sound self-evident, *if* we are still talking about selfishness, altruism and welfare in the biological sense because many things we do would count as *biologically selfish* even though they clearly are not *morally selfish*. Am I behaving selfishly if a maniac tries to kill my family and me and I defend them and myself to the extent that this person is killed? Or am I behaving selfishly if I accept a well-paid job that John is also interested in, and as a consequence am able to financially support a larger family than John can? Or am I behaving selfishly if I park my car in a proper way in the car park but somebody else happens to crash into it by accident and as a result of their injuries is unable to have any children? Or am I behaving selfishly if I go to the library and borrow the only book they have on artificial insemination and this action of mine prevents another person from knowing about this method of becoming pregnant and as a consequence she fails to have any children of her own? The answer is 'yes' if we are talking about biological selfishness because in all of these cases I decrease another person's reproductive fitness and increase mine, and 'no' if we have in mind moral selfishness because I am not doing anything morally wrong. Hence, it is possible for people at one and the same time to behave selfishly in the biological sense but still not be morally selfish.

Note also that this shows that it is not self-interest *per se* that makes a person selfish. If there is no natural way for my wife and me to have children, then it is in our self-interest to try to find out if there are any artificial options available. Hence I am acting out of self-interest when I go and borrow books about artificial insemination, but there is nothing selfish about that. It becomes a selfish act when I pilfer all books about artificial insemination from the public libraries in town, refusing to share these scientific insights with other people. Thus, the definition I offered was not 'People are morally selfish if they are motivated to satisfy their own self-interest', but 'People are morally selfish if they are motivated to satisfy their own self-interest *without regard for others.*'

Hence 'morally selfish' and 'morally altruistic' are mutually exclusive, but they are not exhaustive concepts. That is to say, actions that are morally selfish cannot at the same time be morally altruistic, but it is still possible for behaviour to be *neither* selfish *nor* altruistic. It is neither selfish nor altruistic as such to eat food, to find shelter, to protect one's children, to watch TV, to buy a car, to get a job or to do science. In fact, most of the actions we do probably belong to this third category. People become morally selfish when they exceed what we might call the bounds of 'legitimate self-interest' or 'legitimate self-realization', and start doing things which show disrespect for others and their interests. They become greedy, start to treat other people unfairly and so on. Or to take a more trivial matter as an illustration. It is in my legitimate self-interest to watch TV. But if I am the one in my family who always decides what TV programme we should watch, then I am starting to behave selfishly because I have not taken into consideration the legitimate self-interest of the other members of my family and that is something that I ought to do. Realizing that 'morally selfish' and 'morally altruistic' are mutually exclusive but not exhaustive concepts also explains why we do not regard a moral norm such as the one Jesus claims to be the second greatest after loving God as an expression of selfishness or egoism, namely, 'Love your neighbour as yourself' (Matt. 22: 39). True self-love is not equivalent to selfishness. It can therefore be combined with loving others. This is true about reciprocal action as well. There is nothing morally selfish about helping each other with certain tasks to the mutual advantage of both. Cooperation is an example of legitimate self-realization.

Consequently, Dawkins' claim that if we want to promote universal love and create a society in which people cooperate generously and unselfishly towards a common good, then we can expect little help from our biological nature, derives its plausibility from a confusion of different senses of the value concepts involved. But it is not merely the case that it is possible for people at the same time to be biologically selfish and morally non-selfish or even altruistic because their behaviour is an expression of legitimate self-interest (the third category). It is also possible for people to be *morally altruistic* and *biologically selfish* at the same time. Suppose people are ready to recycle even at their own inconvenience and their motive for these actions is concern for the environment and for future generations. Because the motive behind their actions is helping others even at their own inconvenience, they are behaving morally altruistically. In this way they 'cooperate generously and unselfishly towards a common good', a sustainable society. Suppose now that Dawkins can show that this is really biological selfishness because the effect of this behaviour is the production of more offspring in the following generations. The biological consequence of these actions is that more copies of these people's genes get transmitted into future generations. This behaviour is still morally altruistic because of their motive and own inconvenience, and these people can still 'cooperate generously and unselfishly towards a common good'. We can be morally altruistic and biologically selfish at the same time.

These are possibilities Dawkins completely overlooks simply because he conflates these different senses of selfishness and altruism and related notions such as universal love and generosity. The result is sheer confusion and a serious misconception of the relevance of evolutionary theory for society and ethics.

If the sociobiological extension of evolutionary theory to human behaviour is correct, the kind of behaviour that we (to quote Dawkins again) 'can expect little help from our biological nature' to perform if we want to is instead of the following sort. Suppose some of the people who have a concern for the environment do not have any children, nor do they intend to have any children, and the same is true of their close relatives (or, at any rate, let us assume that these people have no contact with their relatives), and nevertheless they recycle. They tell us that their motive for recycling is concern for other living things and future human generations, and they perform these actions simply because they think they are morally right.

But if this is possible, then, of course, the truth that fills Dawkins 'with astonishment', namely, that 'we are survival machines – robot vehicles blindly programmed to preserve the selfish molecules known as genes' is false, because when it comes to at least these people, they themselves are in charge and not their genes (Dawkins 1989: v). (We can image that the genes, figuratively speaking, are begging these people to please have children and not to be so selfish as thinking merely of themselves and their interests, but they refuse to listen to their genes.) However, not even Dawkins himself really believes this 'truth' which fills him with astonishment because a couple of pages later he writes, 'Our genes may instruct us to be selfish, but we are not necessarily compelled to obey them all our lives' (Dawkins 1989: 3). These claims are clearly contradictory and cannot be true at the same time (unless, of course, biologists mean something very different by 'machines' than we normally do).

It is also worth noticing that Dawkins' worries about the fact that universal love and the welfare of the species as a whole are concepts that simply do not make evolutionary sense are, of course, only a problem for a *scientistic* biologist, who thinks that evolutionary theory alone could and should explain moral behaviour (Dawkins 1989: 2).

The same kind of conflation of these different senses of altruism and selfishness can also be found in the writings of Wilson and Alexander. Recall Wilson's explanation of Mother Teresa's behaviour. Even though Wilson uses the term 'altruism' to cover cases in which animals (including humans) act so as to promote the fitness of another animal at some costs in fitness to themselves, his explanation of Mother Teresa's behaviour is that she is selfish because she as a Christian expects to be rewarded in heaven. But if this is true (one cannot help but wonder how Wilson can know that helping others in need is not reward enough) she would have been morally selfish because of a bad *motive*. But Mother Teresa could still have been *biologically* altruistic if her behaviour resulted in the increase of another person's chances to survive and reproduce to the disadvantage of her own reproductive fitness. This appears in fact to be the case. Theresa had no offspring herself and she did not care in particular about her own relatives' offspring but spent most of her time helping poor and sick strangers living in Calcutta, so that they in fact could increase their reproductive fitness. Consequently, it is even possible for people to behave *biologically altruistically* but still be *morally selfish*.

To sum up, we have seen that it is possible to combine the moral and biological use of these terms in different ways: human behaviour can be both biologically selfish and morally selfish, biologically altruistic but morally selfish, biologically selfish but morally altruistic, or both biologically altruistic and morally altruistic.

With these clarifications in mind let us once again take a look at Alexander's remarks about the 'greatest intellectual advance of the century' and its effect on 'humanity's view of itself' (Alexander 1987: 3). One such effect he points out is that we can understand morality only if we see 'society as a collection of individuals seeking their own self-interests'. This idea 'runs contrary to what people have believed and been taught about morality'. It contradicts, for instance, the belief which nearly all humans share that 'it is a normal part of the functioning of every human individual now and then to assist someone else in the realization of that person's own interest to the actual net expense of those of the altruist'. But evolutionary theory tells us that there is 'not a shred of evidence to support this view of beneficence'. So we have to 'start all over again to describe and understand ourselves, in terms alien to our intuitions', namely that we are programmed to be selfish.

In what way does Alexander use the words 'self-interest', 'altruist' and 'beneficence' in these sentences? I take it that what people normally assume is that they themselves and other people sometimes behave morally altruistically and that their actions are sometimes motivated by the desire to benefit others even at their own expanse. This is probably what Alexander refers to when he writes, 'it is a normal part of the functioning of every human individual now and then to assist someone else in the realization of that person's own interest to the actual net expense of those of the altruist.' But if we accept this interpretation then it becomes puzzling why Alexander thinks the fact that evolutionary theory is unable to find evidence to support such behaviour would undermine the existence of moral altruism, since *evolutionary theory does not after all study intentions and motives*. Furthermore, even if Alexander were able to show that the apparently altruistic behaviour, which is a normal part of the functioning of every human individual, is in fact biologically selfish, this is still compatible with the very same behaviour being morally altruistic, as we have seen.

On the other hand, if we interpret the word 'altruist' as referring to a biological altruist in the sentence, 'it is a normal part of the functioning of every human individual now and then to assist someone else in the realization of that person's own interest to the actual net expense of those of the altruist', then it does not make sense. This is because people in general do not know much, if indeed anything, about biological altruism – something which Alexander cannot reasonably have failed to notice. The upshot is that either (a) Alexander talks in this paragraph about *biological* altruism and selfishness in which case his ideas about a dramatic change in self-view seems clearly misplaced or (b) he talks about *moral* altruism and selfishness and his ideas about a dramatic change in self-view makes sense, but then he cannot find support for these ideas in evolutionary theory.

What these evolutionary biologists in fact do is to take a *heavily value-loaded* term from culture – 'selfishness' – redefine it and apply it to nature, which might be problematic enough, but then re-apply it to culture but now with a different meaning than the word normally has when used to evaluate human behaviour, and then to make things even worse, to conflate these two senses of selfishness. The result is sheer confusion. Moreover, a standard procedure in much social science is to try to avoid, if possible, heavily value-loaded terms when explaining human behaviour. If these biologists want to be good social scientists, there are many things they can learn from social scientists, and this is clearly one such thing.

There is one more important reason why it is desirable that evolutionary biologists develop a less value-loaded terminology to express the content of the gene selection theory. I think a lot of people are not able to say what conditions must be satisfied for a behaviour to count as morally selfish, but they at least know that it is a behaviour that is bad and blameworthy. The person who has been behaving selfishly has done something *wrong* or *bad* and accordingly is *blameworthy* by way of deserving the disapproval of others and the reproach of his or her own conscience. Selfish behaviour is a behaviour that deserves condemnation. This is something they know for sure even if they would not agree on exactly what types of behaviour satisfy this criterion. Nevertheless they certainly can discriminate, as Ashmore points out, between those who devote their lives to caring for lepers and those whose career is loan-sharking, and between the motive of a parent assiduously attending to a sick child and the motive of one who beats a child that disturbs his sleep.

But in this sense genes cannot reasonably be said to be selfish. Genes cannot be *blamed* for behaving as they do. I am not suggesting, of course, that these evolutionary biologists are unaware of this, but what makes one wonder is why, despite this knowledge, they still choose to use this heavily value-loaded term in explaining the behaviour of genes and humans. If, as Rolston suggests, they use 'self-preservation', 'self-actualization' or similar words, they would, of course, not sell nearly as many books nor attract so much publicity, but if they take science seriously that is a price they should be willing to pay (Rolston 1999: 84f). The behaviour of genes would then not be selfish but 'self-preserving'. Organisms would 'defend' their lives in both competition and symbiosis with other organisms of the same or other species. This makes much more sense and we avoid a pejorative term and the possibility of misconceptions by conflating different senses of key terms. Dawkins for instance would then write,

> Humans and baboons have evolved by natural selection. If you look at the way natural selection works, it seems to follow that anything that has evolved by natural selection should be *self-preserving*. Therefore we must expect that when we go and look at the behaviour of baboons, humans, and all other living creatures, we will find it to be *self-preserving*,

and not like he is doing now, namely,

> Humans and baboons have evolved by natural selection. If you look at the way natural selection works, it seems to follow that anything that has evolved by natural selection should be *selfish*. Therefore we must expect that when we go and look at the behaviour of baboons, humans, and all other living creatures, we will find it to be *selfish*.
> (Dawkins 1989: 4, emphasis added)

Rolston also points out that it seems to be very hard indeed for a gene to be selfish in the biological sense if one must be split in half at very reproduction, and when a gene has fitness only in the company of other genes in an organism (Rolston 1999: 72, 96).

To sum up, we have seen that there are good reasons for distinguishing between what I called the gene selection theory and the selfish gene theory and also for rejecting the latter. In particular, evolutionary theory does not undermine the view common in moral philosophy and theology that morality is *not* about selfishness or inclusive fitness. This is so because both biological altruism and selfishness are compatible with either moral altruism or selfishness. The impression that evolutionary theory does undermine the ordinary conception of morality is caused by the fact that these biologists in their argument conflate different conceptions of self-interest, selfishness and altruism.

Is the objectivity of morality an illusion?

Let us consider the second consequence that these biologists think science applied to human behaviour entails: is it the case that if we accept evolutionary theory then we have to accept also that the objectivity of morality is an illusion? Ruse and Wilson maintain that 'the evolutionary explanation makes the objective morality redundant' (Ruse and Wilson 1986: 187). They write, 'We believe that implicit in the scientific interpretation of moral behaviour is a conclusion of central importance to philosophy, namely that there can be no genuinely objective external ethical premises' (Ruse and Wilson 1986: 186). But from what premises more exactly does this conclusion follow and what is meant by 'objectivity' in this case? This is less clear. It seems that the idea is that *if* one accept that 'morality, or more strictly our belief in morality, is merely an adaptation put in place to further our reproductive ends', *then* it follows that there are no genuinely external ethical premises (Ruse and Wilson 1993: 310). *If* 'ethics ... is an illusion fobbed off on us by our genes to get us to cooperate', *then* there cannot be any objectivity in moral reasoning (310). *If* 'ethical premises are the peculiar products of genetic history, and they can be understood solely as mechanisms that are adaptive for the species that possess them', *then* it follows that 'the ethical code of one species cannot be translated into that of another' (Ruse and Wilson 1986: 186).

These biologists can also explain, as we have seen, why we tend to think that morality is objective. The explanation is that 'human beings function better if they are deceived by their genes into thinking that there is a disinterested objective morality binding upon them, which all should obey' (Ruse and Wilson 1986: 179). So in fact it is our genes that fool us into believing, for instance, that 'Do to others as you would have them do to you' is objective and thus ought to apply to all people. But although evolution leads us to believe that there is a difference between right and wrong, there really is no such thing. Consequently, if this revolutionary implication of the evolutionary theory becomes well known, people will cease to 'function better' when they realize that ethics is merely an illusion fobbed off on them by their genes to get them to cooperate.

Although Ruse and Wilson never explicitly address what they mean by 'objective morality' or 'objective ethical premises', these quotations give us a good hint. Morality cannot be objective in the sense of applying to all people (valid in an intersubjective way) because it is not the outcome of divine authority or reason or the like, but an illusion fobbed off on us by our genes to merely further our reproductive ends.

There are, however, several problems with arguing in the way Ruse and Wilson do. One difficulty, of course, is that the conclusion presupposes that the evolutionary explanation of morality is true, something we have seen that there are reasons to doubt. But the argument in fact presupposes something more, namely a particular form of Scientism. Only if we assume that the evolutionary explanation of morality is not merely true but also *complete*, do we have reasons to think that the objectivity of morality is an illusion. But then we need an argument that can take us from 'Morality is an adaptation' to

'Morality is *merely* an adaptation', and from 'Ethical premises can be understood as mechanisms that are adaptive for the species that possess them' to 'Ethical premises can *solely* be understood as mechanisms that are adaptive for the species that possess them.'

No such argument is given, however. It seems rather that they simply presuppose the truth of T1, that the only kind of knowledge we can have is scientific knowledge and T2, that the only things that exist are the ones science can discover. But T1 and T2 are claims, as we have seen in Chapter 2, which cannot be scientifically justified and there are also good reasons to believe them to be false.

Be that as it may, their argument for the claim that the objectivity of morality is an illusion still fails because it is *self-refuting*. It is maintained that ethical norms or beliefs cannot be objective because they are merely the product of evolution. They are rather an adaptation put in place to further our reproductive ends and nothing else. This argument seems to be as follows:

(1) The ethical beliefs we have are merely the product of evolution.
(2) The most dominant determinant in evolution is maximizing fitness, that is, the production of the most offspring in the next-following generations.
(3) Therefore, ethical beliefs are not objective or universal, but are solely an adaptation put in place to further our reproductive ends.

But what is not the product of evolution in this sense? Is not science or scientific theory also a biological product? It certainly seems to be the case. Scientific thinking, like moral thinking, is all done on circuits in the brain that the genes have made. It is essentially the same genetic make-up that is being used in both cases. Thus science and not merely ethics can and ought to be Darwinized or biologized. This is something at least Ruse willingly admits because he maintains, 'the principles of scientific reasoning or methodology … have their being and only justification in their Darwinian value, that is in their adaptive worth to us humans … I argue that the principles which guide and mould science are rooted in our biology, as mediated by our epigenetic rules' (Ruse 1998: 155).

But the problem is that by using the same logic as in the argument above we can also obtain the conclusion that scientific theories like evolutionary theory are not objective or true but are merely adaptations put in place to further our reproductive ends. Just as ethics is an illusion fobbed off on us by our genes to get us to cooperate, so is science! Consequently and by implication, the only reason why Ruse and Wilson put forward this argument about the non-objectivity of morality (even if they do not know it themselves) is that they (or rather their genes) want to maximize their own genetic fitness. What their scientific theorizing is all about is increasing their offspring in the next generations. But then, of course, we have no reason to believe that their claims about morality are true. In fact, if these biologists are right that we are survival machines for our genes (that is, that our behaviour is firmly under the control of the genes), then it is even very, very unlikely that Alexander, Dawkins, Ruse, Wilson or anyone else would be able to detect that we are deceived by our genes into thinking that there is a disinterested objective morality binding upon them, which all should obey. Should we really think our 'selfish' genes would 'want' us to come to know this? Hardly. Our behaviour is after all firmly under the control of the genes. So if these biologists are right, then they are wrong. Their denial of the objectivity of morality is self-refuting.

This brings us back to the question of the adequacy of the evolutionary explanation of ethics and more broadly human behaviour. If these biologists want to avoid maintaining a self-refuting theory, they have to allow that there are in human behaviour important exceptions to the evolutionary explanation. To the extent that these scientists in formulating their own research programme can escape from the bondage of their own theory and be engaged in a rational inquiry about the status of a scientific theory, it seems that philosophers and theologians can also claim that they are engaged in rational selection and justification of ethical norms and not in natural selection and biological reproduction.

Let us conclude by saying that it is likely that evolutionary theory can be of relevance for morality and moral reasoning in that it can provide us with new information that can undermine ethical norms and beliefs. Although this is very important, it is nevertheless true that almost any scientific discipline can be of relevance for ethics in this sense. This by itself does not warrant any grand claims about a biologization of ethics. However, two of the claims made by evolutionary biologists who belong to this category, if true, seem to have profound implications for moral behaviour and moral reasoning. Those are that morality is ultimately a matter of selfishness, and that the objectivity of morality is an illusion. We have seen, however, that the first expresses a conceptual confusion and the second is self-refuting.

Scientific evolutionary ethics

I have tried to assess in what way evolutionary theory can be of significance for ethics. One way it can be relevant is, as we have seen, that it can undermine existing ethical theories, norms or beliefs. But there is of course another, more direct way that evolutionary theory can be of great significance for ethics, and that is that it can replace traditional ethics with a scientific evolutionary ethics. The idea is that evolutionary theory can itself justify ethical norms and beliefs and provide us with a new ethic. (Note, however, that not all of the biologists that claim that evolutionary theory is of great relevance for morality think this is possible. Alexander writes, 'I am convinced that biology can never offer ... easy or direct answers to the questions of what is right or wrong' and Dawkins maintains that he is 'not advocating a morality based on evolution' (Alexander 1987: xvi and Dawkins 1989: 2).

Ruse and Wilson are the key advocates of the view that evolutionary theory can tell us what we ought to do because of the possibility of discovering 'ethical premises inherent in man's biological nature' (Wilson 1978: 5). Ruse and Wilson maintain that 'it will prove possible to proceed from a knowledge of the material basis of moral feeling to generally accepted rules of conduct. To do so will be to escape – not a minute too soon – from the debilitating absolute distinction between *is* and *ought*' (Ruse and Wilson 1986: 174). Evolutionary theory can establish certain moral values because we can obtain moral values from knowledge of the biological facts. In other words, a presupposition for the success of an evolutionary ethic is a successful refutation of the naturalistic fallacy, that is, the idea that one commits a logical error if one tries to deduce ought-statements from exclusively is-statements.

In his critical discussion of the naturalistic fallacy Wilson takes John Rawls' theory of justice as an example of the mistake philosophers tend to commit. He writes that Rawls suggests that

> justice be defined as fairness, which is to be accepted as an intrinsic good. It is the imperative we would follow if we had no starting information about our own future status in life. But in making such a suggestion Rawls ventured no thought on where the human brain comes from or how it works. He offered no evidence that justice-as-fairness is consistent with human nature, hence practicable as a blanket premise. ... Had Kant, Moore, and Rawls known modern biology and experimental psychology, they might well not have reasoned as they did. [But] countless scholars in the social sciences and the humanities ... like Moore and Rawls, have chosen to insulate their thinking from the natural sciences. Many philosophers will respond by saying, ethicists don't need that kind of information. You really can't pass from *is* to *ought*. You can't describe a genetic predisposition and suppose that because it is part of human nature, it is somehow transformed into an ethical precept.
>
> (Wilson 1998: 56–7)

This Wilson rejects. The reason is that to

> translate *is* into *ought* makes sense if we attend the objective meaning of ethical precepts. ... They are more likely to be products of the brain and the culture. From the consilient perspective of the natural sciences, they [the ethical precepts] are no more than principles of the social contract hardened into rules and dictates.
>
> (Wilson 1998: 57)

Thus, '*ought* is just shorthand for one kind of factual statement, a word that denotes what society first chose (or was coerced) to do, and then codified' (58).

Wilson's argument here is not as clear as it might be. First, it is true, as Wilson writes, that Rawls presupposes that 'justice-as-fairness is consistent with human nature'. Rawls takes for granted that it is within the powers of human beings to fulfil the demands of justice-as-fairness. If biology can establish that that is not the case, then biology has in a very effective way undermined the possibility of this kind of theory of justice. But to argue in this way is not to reject the naturalistic fallacy. It is rather to make use of a norm widely accepted among philosophers, namely that *ought implies can*. If a person ought to do a certain action then it must be possible for that person to perform this action. So if our biology makes it impossible to treat other people in a fair way then we cannot blame ourselves for this. (In a similar way, I cannot be blamed for not jumping into the water and trying to save a drowning person if I cannot swim myself.) Thus, if we accept that ought implies can, then scientific results can directly entail that certain ethical theories or norms are unacceptable. Hence, if human behaviour is firmly under the control of the genes and (T) is true (that is, that the most dominant determinant in human behaviour is maximizing fitness), then we might be genetically predisposed in such a way that it is impossible for us to eliminate inequalities and strive towards justice-as-fairness.

We have, however, already seen that there appear to be good reasons to question (T) and thus also the explanatory scheme (E). However, the important point here is that since the 'ought implies can' relationship is widely known and recognized among moral philosophers, it cannot be the case that philosophers have chosen to insulate their thinking from the natural sciences in this sense. Hence, Wilson in his criticism of Rawls confuses the naturalistic fallacy and the ought-implies-can principle.

But Wilson is also saying something else, which sounds more like an argument against the naturalistic fallacy. He tells us that ethical precepts are from the perspective of the natural sciences no more than principles of social contract hardened into rules, which means that ought-statements are just shorthand for one kind of is-statement. Since science deals with is-statements, the question is now merely to relate different types of is-statements to each other, and in such a way evolutionary biologists can show that the naturalistic fallacy is itself a fallacy. The idea seems to be that evolutionary theory can tell us what to value and how we ought to behave because we can translate ethical statements into factual statements. Since this is possible, science alone can replace traditional ethics and tell us how to behave morally.

Let us see if this can successfully be done. Suppose we start with the following factual statement,

(1)   Thousands of women are being abused and oppressed by men.

Add now an ethical ought-statement that is common in our present society, a statement expressing an ethical belief that goes against a practice of discrimination once widely accepted:

(2)   It is morally wrong to abuse and directly oppress other people because of their sex, skin colour or ethnic background.

If these premises are true we can conclude:

(3)   Therefore, these men ought not to abuse and oppress these women.

Now, if I understand Wilson right, he thinks that we can validly obtain this conclusion or another ought-conclusion by transforming the ought-statement expressed in premise (2) into a factual claim – and (as he tells us) from the perspective of the natural sciences, (2) is after all merely an expression of what society chose and then codified. Thus,

(4)   In our society today it is considered morally wrong to abuse and directly oppress other people because of their sex, skin colour or ethnic background.

Perhaps we also should add some relevant facts about our biological nature that the evolutionary theory supports us with, such as,

(5)   There are genetic differences that make women less assertive and less physically aggressive than men.
(6)   Humans are genetically disposed to adjust their behaviour to maximize their own genetic fitness.

Recall that Wilson's claim is not merely that evolutionary theory can *explain* why we have changed our moral conviction to (4) from

(7)   In pre-modern society it was often not considered morally wrong to abuse and directly oppress other people because of their sex, skin colour or ethnic background

but that evolutionary theory can also tell us whether this is how we morally *ought to behave*. But how could this possibly be done in the way that Wilson suggests? Claims such as (4) and (7) are important, but they merely report the kind of moral convictions people at a given time possess, and claims such as (5) and (6) merely explain why it might be difficult or easy for people to live in accordance with certain moral convictions. Neither group of is-statements can either individually or jointly justify the claim that such behaviour is morally acceptable or unacceptable: how men ought to morally behave towards women. These facts do not tell us what we ought to value or how we ought to behave. The mere fact that a particular practice like slavery is widely accepted in a society does not automatically make it morally right.

But in what way, then, can biologists take us from facts to values? Let us see if we can obtain any additional information about the way evolutionary theory can tell us how we morally ought to behave by taking a closer look at some of the ethical beliefs Wilson himself thinks are justified by this theory. Although Wilson still believes we have a long way to go before we can obtain all our moral values from our biological knowledge, he writes that his scientific evolutionary ethic leads him already now to accept certain values on this basis, for instance, that (a) we ought to protect 'the cardinal value of the survival of the human genes in the form of a common pool over generations'; (b) we ought to 'favor diversity in the gene pool as a cardinal value'; (c) we ought to regard 'universal human rights ... as a third primary value'; and (d) we ought to 'Love the organisms for themselves first' (Wilson 1978: 197–9; 1994: 191).

Let us consider his arguments for (a) and (c). One cardinal value of the new biology of ethics that Wilson identifies is 'the survival of human genes in the form of a common pool over generations' (Wilson 1978: 196). His argument for why we should accept this value as a cardinal value is essentially the following:

> Few persons realize the true consequences of the dissolving action of sexual reproduction and the corresponding unimportance of "line" of descent. The DNA of an individual is made up of about equal contributions of all the ancestors in any given generation, and it will be divided about equally among all descendants at any future moment.
>
> (Wilson 1978: 197)

Again Wilson thinks he can justify a moral value by referring to the biological knowledge evolutionary theory provides us with about human nature and its evolution. The argument seems to be:

(1) Our DNA was once distributed among millions of other humans.
(2) Our DNA will again in the future be distributed among millions of other humans.
(3) Therefore, we should be concerned about the survival of the human gene pool.

But why should we – even if we know that (1) and (2) are true – be concerned about the survival of the human gene pool? In themselves, these facts have no moral direction. How can they alone tell us what to value? The answer must be that they cannot. Rather (3) follows only if we add some additional premise such as,

(4) A rational person ought to be concerned about the survival of his or her DNA.

or

(5) Human beings cannot do anything other than be concerned about the future of their DNA.

If Wilson presupposes the truth of (5), he has not shown how one can successfully infer ought-statements from is-statements, but has again appealed to the ought-implies-can principle. We ought not to do something that we cannot do because of our biological make up. But then, of course, Wilson need not to be worried about the fulfilment of this value because we cannot do anything else than honour it. If Wilson, on the other hand, presupposes the truth of (4), which is an ought-statement, then he has not derived an ought-statement from only is-statements, and he has not given us a reason why we ought to be concerned about the future of our DNA. Why is the future of our DNA so important?

The result would not be any better if we, in the manner Wilson suggests, transform (4) into an is-statement, that is:

(4′) In our society people are concerned about the future of their DNA.

This would not help because I might still be entirely indifferent to the fate of my genes, and, more importantly, wish a justification why this value and not some other value ought to be a cardinal value. If I am not bound to foster the survival of my genes, that is, if (5) is false, information about my genes and what other people think of them cannot settle the issue, because I, and not my genes, am then making the decision.

Once again we come back to the question of the adequacy of the evolutionary explanation of human (moral) behaviour. If the explanation is true, then it seems that *evolutionary theory does not so much give us a new ethic as show us that we cannot actually behave in such a way that is contrary to the goal of maximizing genetic fitness*. The formula is:

> Biology has established that it is not possible for us to do X; therefore we ought not to do X.

But then we do not really need an ethics of biology or any ethics at all because our genes 'tell' us what to do. Or if we need an ethics, it is only in those situations in life in which the survival of our genes is of no relevance; and if we are to believe these biologists, such situations are rare.

Can Wilson's argument for seeing universal human rights as a cardinal value change these conclusions in any way? He writes,

> Universal human rights might properly be regarded as a third primary value. The idea is not general; it is largely the invention of recent European-American civilization. I suggest that we will want to give it primary status not because it is a divine ordinance ... or through obedience to an abstract principle of unknown extraneous origin, but because we are mammals. Our societies are based on the mammalian plan: the individual strives for personal reproductive success foremost and that of his immediate kin secondly; further grudging cooperation represents a compromise struck in order to enjoy the benefits of group membership. ... We will accede to universal rights because power is too

fluid in advanced technological societies to circumvent this mammalian imperative; the long-term consequences of inequity will always be visibly dangerous to its temporary beneficiaries. I suggest that this is the true reason for the universal rights movement and that an understanding of its raw biological causation will be more compelling in the end than any rationalization contrived by culture to reinforce and euphemize it.

(Wilson 1978: 198–9)

The reason why universal human rights ought to be understood as a cardinal value is because of who we are, that we – as biology has discovered – are mammals. Biology has also discovered that the dominant determinant of the behaviour of mammals is maximizing fitness, primarily individual fitness, secondly, genetic fitness and, when necessary, securing individual or genetic fitness through reciprocal behaviour (that is, the *mammalian plan*). Wilson then adds some facts about modern society that make it different from previous human societies and the societies of other animals. This is crucial because even though all of these animals are mammals, these previous human societies did not respect human rights and within these non-human societies we can find nothing equivalent to human rights regulating the behaviour of lions, elephants and chimpanzees. These additional facts are that today's societies are technologically advanced and that the long-term consequences of inequity within such societies will always be visibly dangerous to its temporary beneficiaries. Therefore, universal human rights ought to be respected and regarded as a cardinal value.

Thus the argument looks, roughly, as follows:

(1) Humans are mammals.
(2) The dominant determinant of the behaviour of mammals is maximizing fitness.
(3) Modern societies are technologically advanced societies.
(4) The long-term consequences of inequity within technologically advanced societies will always be visibly dangerous to its temporary beneficiaries and thereby threaten their fitness.
(5) Therefore, universal human rights ought to be respected and regarded as a cardinal value.

The first problem with this argument is that its conclusion is stated too generally. In non-technologically advanced societies the consequences of inequity will not always be visibly dangerous to its temporary beneficiaries. Hence, people in so-called indigenous societies can (if they want to) continue to violate human rights because Wilson does not give a reason why *they* should behave towards other humans differently. Perhaps the best way to maximize fitness in low technological societies is typically to respect merely one's own and perhaps one's relatives' rights. So the universal rights cannot in an evolutionary ethic be as universal as some of us (including Wilson I presume) would like them to be. The proper conclusion is then:

(5′) Therefore, universal human rights ought to be respected and regarded as a cardinal value *within* technologically advanced societies.

But then, if things (the facts) had been different, it seems as if Wilson would allow us to draw a quite different conclusion than (5′). Suppose

(4′) The long-term consequences of *equity* within technologically advanced societies will always be visibly dangerous to its temporary beneficiaries and thereby threaten their fitness

were true, then we can conclude,

(6) Therefore, universal human rights ought *not* to be respected and regarded as a cardinal value within technologically advanced societies.

But if (4′) were true, why should we stop believing that equity is what we should strive to obtain? Is that not a risk worth taking? Even if we would add that our biological nature would not really like to cooperate, should we not try our best to cultivate the moral virtues necessary for equity? At best the evolutionary ethic seems to be able to tell us that if (4) is true then rejecting human rights will have its price, and correspondingly, if (4′) is true then honouring human rights will have its price. This ethic does not give us the means to evaluate whether the price of equity or inequality is worth paying.

Furthermore, is it the case that the factual claim expressed in premise (4) is supported by the evidence? Wilson merely tells us that this is the case, but he does not give us any references to scientific studies that support the idea that the long-term consequences of inequity within technologically advanced societies will always be visibly dangerous to its temporary beneficiaries. This factual claim can certainly be disputed. Even if we do not name them I am certain that most of us upon reflection can think of technologically advanced countries where the temporary beneficiaries of inequity do not see the dangers of not respecting universal human rights. The extent to which these universal human rights are violated in these societies is also relevant to the argument.

Wilson seems, moreover, to get things back to front in his argument for universal human rights. He writes as if the avoidance of these dangers is the main objective and that the *means* to obtain that goal is to honour universal human rights. Universal human rights is the idea that we adopt as a means of avoiding such dangers to our well-being or perhaps to the goal of maximizing our genes in the next generation. We ought to accept universal human rights because otherwise we (or our genes) will be exposed to danger, at least in the long run.

Thus one possible explanation why Wilson gives derivative status to universal human rights may be that he tacitly assumes that the mammalian plan also *ought* to be our imperative. If this is correct then the hidden cardinal value of evolutionary ethic would be:

(7) Rational people follow their biological nature and try to maximize fitness, foremost individual fitness, secondly, genetic fitness, and when necessary secure individual and genetic fitness through reciprocal behaviour (the *mammalian imperative*).

We ought to honour universal human rights *because* in technologically advanced societies the mammalian imperative is best obtained by respecting such rights. But of course neither (7) nor (4) would provide individuals with a compelling reason to respect human rights, if they thought they could get away with not respecting human rights. Those who are primarily interested in maximizing their fitness might respond to Wilson's argument, as Singer points out, either by following Wilson's advice, or by finding some new strategy for controlling power in technologically advanced societies so as to eliminate the danger to them of denying other people their human rights (Singer 1993: 320). Therefore, Wilson is not offering us much of a moral justification for universal human rights at all.

Perhaps it is unfair to Wilson to claim that he tacitly assumes the acceptance of the mammalian imperative in his argument and more generally in his evolutionary ethic, even though it would explain why Wilson gives derivative status to human rights in his argument. But it seems, nevertheless, that a *scientistic* interpretation of the evolutionary theory might lead scientists in that direction. Consider, for example, Herbert Simon's theory about human docility and limited rationality, developed as he tells us within the framework of neo-Darwinism to explain the spreading of altruistic behaviour. By 'docility' he means the tendency we can find among humans to accept social influence. He writes, 'Docile persons tend to learn and believe what they perceive others in the society want them to learn and believe' (Simon 1990: 1666). His idea is that 'because of the limits of human rationality, fitness can be enhanced by docility that induces individuals often to adopt culturally transmitted behavior without independent evaluation of their contribution to personal fitness' (1665). 'Because of bounded rationality, the docile individual will often be unable to distinguish socially prescribed behavior that contributes to fitness from altruistic behavior' (1667). For these and some other reasons Simon concludes that docile persons will necessarily also behave altruistically.

But then does this not mean that the *really rational* way to behave is to try to increase one's personal fitness? Is not the assumption underlining Simon's reasoning that if people were smarter (and perhaps a little less docile) they would be able to screen this culturally transmitted altruistic behaviour and instead behave in a (non-bounded) truly rational way, that is, trying to implement the mammalian imperative? He seems to be very close to accepting the idea that neo-Darwinism entails a morality (or a practical rationality) in which (7) is the cardinal value. But again why think that altruistic behaviour is a manifestation of *limited* rationality? Why not instead think that it is a manifestation of *genuine* rationality? Altruistic behaviour, as Alvin Plantinga points out, is at least from a Christian perspective very rational because it reflects the character of God (Plantinga 1996: 371). But I do not think that we have to bring in Christianity here because moral philosophers in general seem to be ready to draw the very same conclusion.

I suggest that the reason why these evolutionary biologists may end up with these conclusions is that they (consciously or unconsciously) interpret evolutionary theory scientistically, they assume that neo-Darwinism is the whole story of morality. *If morality is merely an adaptation put in place to further our reproductive ends, what else can practical rationality be than maximizing personal and genetic fitness?* Given that evolutionary theory is the only adequate explanation, what else can morality be than an illusion fobbed off on us by our genes to get us to cooperate when necessary to secure genetic fitness? But instead of seeing these consequences as indicators of the inadequacy of the account of morality we can find among philosophers and theologians, they can equally well be used to suggest limitations on evolutionary theory and a scientific account of morality.

Let us summarize this discussion by saying that we have seen that Wilson's argument for universal human rights is far from convincing and that it is by no means able to show that the naturalistic fallacy is in fact itself a fallacy. David Hume's insight that an ought-conclusion must have at least one ought-premise still stands firmly in the way of the development of an evolutionary scientific ethic. Discovering that certain behaviour has a biological basis does not justify that kind of behaviour. The project of developing moral norms and justifying ethical beliefs, therefore, does not genuinely fall within the scope of science or more specifically within the scope of evolutionary biology.

In this and the previous chapter we have examined the merits of the sociobiological research programme when applied to morality and moral reasoning. We have clarified and critically assessed T3, that is, the idea that a biology of ethics can be fashioned and a Darwinian explanation of morality can be given that can replace rival explanations found in the social sciences, the humanities and theology. This has been done by distinguishing between three different ways in which it is possible that evolutionary theory can be of relevance to ethics: (A) it can explain the development and maintenance of morality in human life; (B) it can provide us with new information about human life and its environment that can undermine existing ethical theories, norms or beliefs, and (C) it can justify ethical norms or beliefs and provide us with a new ethic.

The plausibility of project (B) and (C) has been the particular focus of this chapter. Project (B) is perhaps the most obvious way in which evolutionary theory can be of significance for ethics. The problem is that philosophers are in general aware that well-informed moral judgments must not merely be based on appropriate ethical norms, but also on relevant factual information. But at least two of the claims made by biological expansionists would, if true, have a profound impact on the conception of morality we find among philosophers and theologians. These are that morality is ultimately a matter of selfishness and that the objectivity of morality is an illusion. I have shown that the first claim is based on a conceptual confusion. These biologists do not clearly distinguish between selfishness and altruism in its moral and in its biological sense and fail to see that human behaviour can be (1) both biologically selfish and morally selfish; (2) biologically altruistic but morally selfish; (3) biologically selfish but morally altruistic; or (4) both biologically altruistic and morally altruistic. The second claim fails because it undermines itself. The argument is that ethical norms or beliefs cannot be objective because they are merely the product of evolution. They are rather an adaptation put in place to further our reproductive ends and nothing else. But from the biological perspective science too is nothing else than a product of evolution. Thus science cannot be objective, but is an adaptation put in place to further our reproductive ends and nothing else. But then there is no reason for us to believe that the objectivity of morality is an illusion because these biologists' claim is merely the product of evolution. In fact, if their theory is true (that our behaviour is firmly under the control of the genes and that we function better if we believe in the objectivity of morality), then it would be very unlikely that these biologists would be able to discover that the objectivity of morality is an illusion. So if these scientific expansionists are right, they are probably wrong.

Project (C) provides the most direct way in which evolutionary theory could be of relevance for ethics. The claim is that evolutionary theory can itself justify ethical norms and beliefs and provide us with a new ethic. However, to be able to do this biological expansionists must deal with the naturalistic fallacy. Wilson thinks that the naturalistic fallacy is something biology can refute because from the perspective of the natural sciences, the ethical precepts are no more than principles of the social contract hardened into rules and dictates. Thus ought-statements are just shorthand for one kind of factual statement and since science deals with factual statements, it can also deal properly with moral statements. But we have seen that when we 'translate' ought-statements into factual statements in the way Wilson suggests, we can still not obtain any ought-conclusion telling us how we ought to behave or what we should do. Consequently, Wilson must get his ought-premises from some domain other than science, and thus the project of developing moral norms and justifying ethical beliefs does not genuinely fall within the scope of science, or more specifically within the scope of evolutionary biology.

CHAPTER FIVE

The scientific explanation of religion

Our focus has been on Scientism and we have seen that it takes a number of different forms. We have critically assessed three of these forms of Scientism. We have found scientists that attempt to expand the boundaries of science, by claiming that (T1) the only kind of knowledge we can have is scientific knowledge, that (T2) the only things that exist are the ones science has access to, and that (T3) science alone can answer our moral questions and explain as well as replace traditional ethics. We now come to a fourth kind of Scientism, what I in Chapter 1 called 'existential Scientism', namely the idea that (T4) science alone can answer our existential questions and explain as well as replace traditional religion. The idea is that science cannot merely explain religion but also in some sense take its place; it can in the terminology of traditional religions, offer us salvation. We can and must put our faith in science or more exactly in scientific materialism (or scientific naturalism). Two key advocates for this kind of view are Richard Dawkins and Edward O. Wilson, and in what follows we shall critically assess their interesting and provocative claims.

On this point Dawkins and Wilson and other scientific expansionists go against the more generally accepted view that metaphysics, ideology and religion fall outside the scope of the sciences. John E. Post, for instance, writes,

> science stops short of metaphysics ... by not inquiring into whether scientific truths represent the real or ultimate nature of the world, assuming there is such a thing, or whether they are just one among many equally privileged kinds of truth. How 'fundamental' are the truths of physics, and in what sense? Do they enjoy some sort of unconditional priority over truths from other domains? In particular, is it a more basic truth that the sunset is a scattering of photons than that it means certain things to us? And what about the fact that all things in the world have various objective physical properties – mass-energy, say or gravitational relation to neighbors, or reaction to some energy field? Does this mean that all things, persons included, are physical in nature – that they are nothing but material things? Many scientists have definite views on these issues, but not in their professional roles as scientists. Discussions of such matters are not to be found in their scientific research papers or textbooks. Nor do scientists in their professional roles discuss the relations between science and religion, whereas metaphysicians traditionally do.

(Post 1991: 3–4)

Furthermore, Bråkenhielm and Hansson write,

> Walter Gilbert ... has ... claimed that the mapping of people's DNA is 'the ultimate answer to the commandment "know thyself".' A research project with the ambition of giving the ultimate answer to the question 'Who is man?' is something more than merely a research project; it is a worldview project. ... It is one thing to determine the basic structure of the DNA-molecule, a task that ought to be left to the biologists, quite another thing to put the result of the biological research in a wider ideological framework and, for instance, make conclusions about man's basic nature or about what is right and wrong in morality.
> (Bråkenhielm and Hansson 1995: 100–11, my translation)

Bråkenhielm, Hansson and Post, all seem to defend the view that we must distinguish between scientific claims, on the one hand, and religious, metaphysical, or ideological claims, on the other, and that scientists in their professional role lack a mandate to engage in religious or ideological debates. Thus, Bråkenhielm, Hansson and Post would all maintain that in the writing of Wilson and Dawkins there is a slide from scientific theory development and justification to decisions on different ideological, moral and religious issues. But, of course, this is not how Wilson and Dawkins see the matter. They do not think it is a question of sliding but a completely legitimate expansion of science (in particular of biology) to new areas and issues.

The question we will have to consider is whether the recent development in biological science, upon which Wilson and Dawkins say they base their arguments, gives us good reason to reconsider scientific restrictionism with regards to religion. Of course, there are many ways in which evolutionary theory can be of relevance for religion, and thus some of their claims might be justified whereas others might not. I suggest that in the same way we characterized three claims about the significance of evolutionary biology to ethics, we, distinguish three claims according to which it is possible that evolutionary biology can be of significance for religion:

(A) Evolutionary biology can *explain* the development and maintenance of religion in human life. It can give the best account of why people behave religiously or why they believe in God or a sacred, transcendent reality.

(B) Evolutionary biology can provide us with new information about human life and its environment that *undermines* (or supports) existing religious beliefs.

(C) Evolutionary biology can *replace* traditional religions and provide us with a new religion, mythology or view of life.

Let us consider the merit of each one of these different ways in which evolutionary theory could be of relevance to religion. In this chapter I will focus on (A) and state the content of the evolutionary explanation of religion and some of its implications, followed by a critical assessment of it.

Thereafter the scientific case for projects (B) and (C) will be stated and also critically evaluated.

The Darwinian account of religion

Wilson maintains that 'we have come to the crucial stage in the history of biology when religion itself is subject to the explanations of the natural sciences' (Wilson 1978: 192). But he also admits that religion constitutes the greatest challenge to evolutionary biology. The reason for this is that 'religion is one of the major categories of behavior undeniably unique to the human species' (175). We cannot find any parallel to it in wild nature. Whereas in morality we can find some similarities between animal behaviour and moral behaviour (for instance in respect to reciprocal cooperation), this is not true when it comes to religious behaviour. There exist no prayers, religious rituals or beliefs in God among members of other species living on this planet. Religion is a truly unique human phenomenon. But the ability to be religious somehow emerged out of nature, where before there was no such ability. Can evolutionary theory explain this phenomenon and in such a way that any other kind of explanation becomes superfluous?

The strategy evolutionary biologists use for explaining religion is the same as the one they use to explain morality. Wilson writes, 'When the gods are served, the Darwinian fitness of the members of the tribe is the ultimate if unrecognized beneficiary' (Wilson 1978: 184). Religion has high survival value in the same way as morality and that is the reason for its existence. Again,

> The highest forms of religious practice, when examined more closely, can be seen to confer biological advantage. Above all they congeal identity. In the midst of the chaotic and potentially disorienting experiences each person undergoes daily, religion classifies him, provides him with unquestioned membership in a group claiming great powers, and by this means gives him a driving purpose in life compatible with his self-interest.
>
> (Wilson 1978: 188)

So the thesis and explanatory scheme are basically the same as before:

(T$_r$) The most dominant determinant in religious behaviour is maximizing fitness, that is, the production of the most offspring in the following generations.

(E$_r$) Religious belief p, religious behaviour q or religious institution A exists and continues to exist because it emerged and continues to function as a strategy (or part of a strategy) adapted to secure the fitness of the individuals and their genes.

(E$_r$) does not attempt to explain merely this or that religious phenomenon, allowing that other religious phenomena might have a different explanation, but rather attempts to explain religion as such. The idea is that beliefs, myths, rituals and the institutional structures of the religions may differ greatly, but this does not matter because the function of religions is ultimately the same, and it is to protect the genes and secure the fitness of the individual.

The question we have to address is whether this explanation is more successful in the case of religion than it was in the case of morality. Again, (T$_r$) and (E$_r$) claim more than what the scientific expansionists can deliver. This is something they are also willing to admit. What Wilson and others attempt to do instead is to assemble more and more examples of religious belief, rituals and activities that can be successfully explained as fitness-maximizing strategies and in this way build an inductive case for the truth or likelihood of (T$_r$) and (E$_r$). This means, however, that (T$_r$) and (E$_r$) can be disconfirmed if we can find examples of religious belief, rituals and activities that cannot successfully be explained as fitness-maximizing strategies. Religious belief and behaviour could exist that do not favour, and may even hinder, the survival and reproduction of the individuals and their genes.

As I said before, once we take seriously that human beings are a part of the evolutionary scheme, we cannot deny the legitimacy of trying to explain why we have certain religious ideas or behave religiously by relating them to our evolutionary history. The crucial question is, however, *how much* of religion can be explained in this way. Can evolutionary theory offer us a complete explanation of religion in such a way that it excludes other scientific as well as theological and philosophical explanations?

It is quite reasonable to think, as Wilson does, that religion contributes to reproductive success. It provides people with an identity and affinity with the group as well as loyalty towards it. Religion sanctions morality and gets people to cooperate for their mutual good and therefore puts people in a better position to leave more offspring than they would have otherwise done. Fertility is a fundamental feature of religion. Religious believers worship the sun, pray for rain, seek cures for diseases and so on. Wilson's conclusion based upon observations like these is that 'elementary religions seek the supernatural for the purely mundane rewards of long life, abundant land and food, the avoidance of physical catastrophes, and defeat of enemies' (Wilson 1975: 561). But he also maintains that the "ultimate, genetic motivation" for religious behaviour is

> probably hidden from the conscious mind, because religion is above all the process by which individuals are persuaded to subordinate their immediate self-interest to the interest of the group. Votaries are expected to make short-term physiological sacrifices for their own long-term genetic gains. Self-deception by shamans and priests perfects their own performance and enhances the deception practiced on their constituents.
>
> (Wilson 1978: 176)

Thus both ordinary believers and their leaders are probably unaware of what is really going on in religion, but religion, like morality, is a way in which the genes have tricked humans to cooperate with others and care about them. While the individual pays, his or her genes and tribe gain. Religion is adaptive. But why does Wilson think that shamans and priests must be self-deceived or more generally that religious believers are unaware of what is really going on in religion (that is, that they are in fact persuaded to subordinate their immediate self-interest to the interest of the group)? The explanation seems to be that Wilson again tacitly presupposes the *mammalian imperative* (p. 75), namely that rational people follow their biological nature and try to maximize fitness, primarily individual fitness, secondly, genetic fitness, and when necessary secure individual and genetic fitness through reciprocal behaviour. If people were a little smarter they would be able to screen religious behaviour and discover what was really going on. They would then as rational individuals drop religion like a hot potato and instead act so as to increase their individual fitness. According to the mammalian imperative, rational people will not want to be persuaded or forced into producing benefits for others at their own expense, especially if they could get away with less. Assuming this and since religious people typically do not cease to be religious, Wilson needs an explanation, and the one he offers is in terms of unawareness and self-deception. So if we presuppose the validity of the mammalian imperative, then Wilson's talk about unawareness and self-deception seems to make sense.

But believers are not unaware that religion offers 'mundane rewards'. In Genesis we can read about Yahweh's promise to Abram:

> I am God Almighty; walk before me and be blameless. I will confirm my covenant between me and you and will greatly increase your numbers. Abram fell face down, and God said to him: "As for me, this is my covenant with you: You will be the father of many nations. No longer will you be called Abram; your name will be Abraham, for I have made you a father of many nations. I will make you very fruitful; I will make nations of you, and kings will come from you.
>
> (Gen. 17: 1–6)

The Israelites believed in this promise and that God would give them a 'land flowing with milk and honey' (Num. 14: 8). These ideas can also be found among believers today in more or less radical forms. A radical version of the mundane rewards theme can be found in the preaching of Kenneth E. Hagin. He tells us that the material prosperity promised in the Old Testament is available even more for Christians: 'We should have super-prosperity and super-success' (Hagin 1978: 16). But, of course, religions are not a unified phenomenon; thus, we can find religious movements that emphasize chastity and poverty rather than fertility and prosperity. There are many different ways in which one can be religious.[1]

Nevertheless, religious believers are often aware that religion can offer mundane rewards. What they would contest is, rather, that this is *all* religion does. That is the crucial point, namely Wilson's claim that religion can be explained as *nothing but* a fitness maximizing strategy. On this point surely religious believers would disagree. They do not believe that religion simply offers fertility. For many of them religion is primarily a response to life, to fertility and to what they experience as a divine presence within or beyond our world. All things are in some way an expression of this ultimate reality, and it can be sensed in the beauty of the world and the awe which it engenders. Some of these people also claim that they have directly encountered this reality, while others have merely experienced glimpses of something transcendent, something beyond limitation and imperfection. Worship and prayer are for them ways of deepening this awareness of the divine and not strategies to secure the production of the most offspring in the next generation, although that could be a spin-off. Furthermore they recognize that ordinary human existence is defective, unsatisfactory or deeply flawed. Therefore, humans need salvation or liberation. Salvation or liberation can be obtained only if present human existence is transformed by what is ultimately real, what many of them call God. On their own account then, humans discover and reflect upon life and fertility, and this evokes religious beliefs and is confirmed by experiences of a divine reality and by the divine making its presence known through, for instance, Holy Scriptures.

From a religious perspective there is therefore nothing surprising if religion offers mundane rewards and is compatible with religious people's own understanding of their religious behaviour. But Wilson's claim that science can explain traditional religion 'as a *wholly* material phenomenon' in the sense that it can establish that religion offers '*purely* mundane rewards' is incompatible with what we can call a *theological explanation* of religion (Wilson 1978: 192; 1975: 561, emphasis added). The theological explanation of the phenomenon of religion is that people actually have experienced a divine presence within or beyond the physical world.

But which one (if any) of the Darwinian and the theological explanations of religion is the correct one? I shall later argue that this is a question that the sciences cannot answer and that we have to turn to philosophy to be able to argue and give reasons why we believe one or the other of these explanations is the correct one. This kind of question is one that falls outside the scope of the sciences but within the scope of the philosophy of religion. What I want to suggest now is rather that a different scientific explanation of religion is possible and is in fact to be preferred over the Darwinian one. Note first that

---

[1] See Cannon (1996) for a discussion of different ways of being religious.

the Darwinian explanation is a functional explanation. It is an explanation in terms of the role religion plays in people's lives. It tells us what kind of job religion does for religious believers.[2] The function religion ultimately serves is as a strategy adapted to secure the fitness of the individuals and their genes. Is there any scientific alternative? Yes, because scientists can offer an explanation that is a parallel to the one offered of the phenomena of morality in the previous chapter.

A non-Darwinian explanation of religion

With the evolution of *Homo sapiens* natural history has reached a point at which certain creatures have evolved who can question their standing in the world in a way that no other part of nature can. It is their self-consciousness that makes this questioning possible. In the interaction with other people and nature, human beings become aware that they are knowers and agents, and thus begin to understand that their beliefs and practice are theirs. Humans realize that their beliefs and practice are, therefore, not necessarily in tune with those of others or with the world, that their path through the world is but one possible path. Thus humans' self-consciousness initiates a process that leads them to attempt to revise and improve their beliefs and to ask questions about which courses of action are for the best and, I would suggest, also about the meaning and significance of their lives. A particular form of being is thus distinctive for humans. Human existence has a concern for itself. As Søren Kierkegaard puts it, the human person not only exists but is 'infinitely interested in existing' (Kierkegaard 1974: 268) and as Martin Heidegger would say, humans are such that their being is in question for them; it is an issue for them (Heidegger 1962). They anticipate death and they sense their finitude. They suffer alienation and search for wholeness, experience guilt and seek forgiveness. In trying to come to grips with these unique human experiences they develop religion. A religion is thus *a strategy to deal with a particular kind of concern, namely what we might call the existential concern.* Religion attempts to liberate from pure self-concern and anxiety and give orientation and courage to be.

Humans encounter experiences of suffering, death, guilt, meaninglessness and the like. In that context they value a practice that makes sense of these experiences, diagnoses them, and helps them find a way through these existential constraints. Religion thus answers questions concerning who we are, why we exist and what the meaning of our life is, and what stance we should take towards experiences of death, suffering, guilt, anxiety, love and

---

[2] See Clarke and Byrne (1993) for a discussion of different kinds of explanations and definitions of religion.

friendship, and the like. This is what religion does, and therefore it exists. In fact, it seems that once a species possesses a certain level of intelligence they have to develop religion (and morality) in contrast to science, which seems to be optional. (Science, as we know it at least, did not arise before the sixteenth century.) A particular species, *Homo sapiens*, on this planet has developed the ability to identify themselves as an 'I' in contrast to other 'selves' and merely physical objects. The members of this species have realized that they exist but could just as well have not existed and that they live in a world that they did not create. Moreover, they have noticed that they can to a significant extent choose what to do and to believe and that other members of their species have chosen other things to do and believe. Therefore, their being has become a question for them. Their lives and their meaning and direction have become an issue. They have an existential concern.

The explanation of religion is thus that humans have developed over evolutionary time outstanding cognitive abilities, an intellect, which includes self-awareness and abstract and critical thinking. It is the possession of this ability or faculty that makes possible the development of not merely science and ethics, but also religion. According to this non-Darwinian explanation of religion, religion has an existential function rather than a fertility function. A religion is thus *whatever people do to come to grips with their existential concern*. It is what people do to handle their human finitude and their awareness of this finitude. Religion in its functional sense is an attempt to solve what it means to exist in the world as a human being.

We can now add to this the characteristic of 'traditional religions' (to use Wilson's terminology) that they typically contain the idea that it is possible for us to overcome existential constraints in a more profound way *only* if we let our lives be transformed by the divine or the sacred, or if we enter into a right relation with the divine or the sacred. The conceptualization of the sacred varies, however, from religion to religion. But what most traditional believers have in common is a consciousness of and trust in a reality beyond the ordinary world, a belief in the existence of a transcendent dimension of reality. Traditional religion thus points to a level of reality that goes beyond the human and the mundane. It is the presence of the sacred that gives the lives of religious believers their substance and meaning – it gives them the strength or wisdom to deal with or get through existential constraints and positively speaking to understand the true significance of love, friendship and virtue. Thus the sacred is also understood to be something of supreme value and, on some accounts, even thought to be the source of all value in the world. In addition the advocates of traditional religion typically believe that without taking into account the sacred a true transformation of our present defective situation is not possible. Therefore, a common concern among traditional religions is to make contact with this transcendent, sacred reality. We can thus say that the specific aim of traditional religion is to help people theoretically understand and practically integrate the sacred dimension of reality into their lives and thereby release its capacity or value for their lives and for their existential concerns.

Traditional religions are, however, not alone in giving an answer to our existential questions; 'secular ideologies' also (to use Wilson's terminology again) offer such answers. Wilson has especially in mind Marxism, but other ideologies in this sense are available, for instance nationalism, feminism, Nazism, existentialism and environmentalism. (These ideologies can, of course, be combined with traditional religions, but – and that is the important point – they can also themselves be a religion in the functional sense.) What scientific expansionists like Dawkins and Wilson now suggest is that science can enter this race and compete with traditional religions and secular ideologies, and that science, in fact, beats them on any point worth attention.

In the next section we will discuss this purported competition, but let us now ask which one of these competing scientific explanations is to be preferred. I shall not attempt to develop a full case in this context, but one way of approaching this question is to ask whether we can find any examples of religious belief, rituals and activities that cannot successfully be explained as fitness-maximizing strategies. Religious behaviour could exist that does not favour, and may even hinder, the survival and reproduction of the individuals and their genes. If this is so and these facts are compatible with the existential explanation then we have a reason for rejecting the Darwinian explanation of religion.

Note first, however, that the problem of self-referential incoherence is a problem scientific expansionists face with respect to their explanation of religion (as well as of morality). Wilson writes that we now have 'the possibility of explaining traditional religion by the mechanistic models of evolutionary biology' and adds, 'If religion, including the dogmatic secular ideologies, can be systematically analyzed and explained as a product of the brain's evolution, its power as an external source of morality will be gone forever' (Wilson 1987: 201). Wilson's argument seems to be:

(1) Traditional religion is a product of the brain's evolution.
(2) The most dominant determinant in evolution is maximizing fitness, that is, the production of the most offspring in the immediately following generations.
(3) Something that maximizes fitness cannot be a valid source of morality.
(4) Therefore, traditional religion's power as an external source of morality will be gone forever once we realize that traditional religion is a product of the brain's evolution.

But again, is science not also a product of the brain's evolution? Yes, indeed it is. Scientific reasoning just like religious reasoning is all done on circuits in the brain that the genes make possible. Thus, science's power as a source of knowledge will be gone forever once we realize that it also is a product of the brain's evolution! Just as religion is an illusion fobbed off on us by our genes to get us to cooperate, so is science. But then, of course, we have no reason to believe the truth of Wilson's argument and of his explanation of traditional religion. Wilson's reasoning about religion is on this point self-refuting. The fact is that the claim that religion is the product of the brain's evolution is in itself not sufficient to warrant any claims about its truth (or lack of it) or its possibility of supporting morality. Evolutionary theory might be true and it might even support a scientific ethic, even though it is unquestionably (like the rest of science) a product of the brain's evolution. The claim that religion is the product of the brain's evolution is moreover compatible with the alternative, existential explanation of religion. A species having an existential concern requires an evolution of the brain in such a way that its development includes the evolvement of self-awareness and reflective and critical thinking. A big brain might even be a necessary condition for the religious quest.

As we have seen, scientific expansionists like Wilson can avoid the problem of self-referential incoherence either by maintaining (T′) that human behaviour is predisposed (but not controlled or determined) by our genes to maximize fitness or (T″) that the most dominant determinant in human behaviour, with the exception of the natural sciences, is maximizing fitness. In our discussion in the last chapter we saw that not merely in the sciences but also in ethics people are able to evaluate beliefs and actions independently of whether these optimize the number of their offspring in the next generation. Since morality and religion have been closely related in human affairs, as these scientific expansionists maintain, we have *prima facie* reason to suspect that in religion as well we can find a capacity to transcend fitness-maximizing concerns.

Let us focus our attention on Christianity (similar accounts can be given for most of the other world religions) to see whether this is the case. It started with one person, Jesus of Nazareth, and he developed his teaching as a reaction to Judaism. Through the preaching of his disciples, other Jews and people of other nations around the Mediterranean Sea came to believe in the teaching of Jesus, and nowadays we can find Christians among most nations of the world, cutting across all ethnic and genetic boundaries. Two key components of the Christian gospel especially relevant in assessing the Darwinian explanation are Jesus' teaching about loving one's neighbour and The Great Commission. Jesus tells his disciples, 'You have heard that it was said, "Love your neighbour and hate your enemy." But I tell you: Love your enemies' (Matt. 5: 43). The command is that those who want to be followers of Jesus ought to be ready to love not merely people of their kin, but also people who do not like them, people who might even persecute them, because they should strive to be 'perfect ... as your heavenly Father is perfect' (Matt. 5: 48). Universal altruism ought to be one virtue Christians should cultivate.

Furthermore, the disciples believed that Jesus had given them the command to 'go and make disciples of all nations, baptising them in the name of the Father and of the Son and of the Holy Spirit, and teaching them to obey everything I have commanded you. And surely I will be with you always, to the very end of the age' (Matt. 28: 19–20). The disciples believed this and preached the Gospel, and people from all around the Mediterranean became followers of Jesus. St Paul summarizes this Christian vision in his letter to the Galatians as follows: 'You are all sons of God through faith in Christ Jesus ... There is neither Jew nor Greek, slave nor free, male nor female, for you are all one in Christ Jesus' (Gal. 3: 26–8).

With the rise of Christianity something therefore happens with tribal morality which Wilson describes by quoting Garrett Hardin: 'The essential characteristic of a tribe is that it should follow a double standard of morality – one kind of behavior for in-group relations, another for out-group' (Wilson 1975: 565). This is the what-you-have-heard part of what Jesus tells his disciples, but it is also what Christians ought not to do; they ought to reject a double standard of morality and adopt an inclusive moral stance. Hence it seems as if Christians have escaped this tribe morality not merely in words, but also in action, at least to the extent that they included in their community both Jew and Greek, slave and freeman, male and female. In other words, Christians did not differentiate between people on the basis of genetic kinship.

But, as Rolston points out, missionary activity of this kind from an evolutionary viewpoint helps to ensure the replication of genes unlike one's own, and that is *not* what the Darwinian explanation predicts would happen (Rolston 1999: 318). If a tribe has a religion that serves their genes well by producing group loyalty and producing numerous offspring, then why would this group attempt to convert people with foreign genes to their religion? If maximizing genetic fitness is the most dominant determinant of religious beliefs and behaviour, then convincing people with foreign genes to become Christians should have been selected against. But this has not been the case. Hence, it cannot be true that 'when the gods are served, the Darwinian fitness of the members of the tribe is the ultimate if unrecognized beneficiary' (Wilson 1978: 184). The existence of world religions provides evidence to the contrary. Religious behaviour is thus not firmly under the control of the genes. A religion such as Christianity cannot plausibly be understood to be merely a strategy adopted to secure the fitness of the individuals and their genes because it transcends genetic boundaries, kinship and group loyalty, and goes universal. Thus Darwinism is unable to explain the universalism that characterizes Christianity as well as most other world religions. Or phrased somewhat more weakly, maximizing fitness cannot be the whole story, and is even far from explaining why traditional religions such as Christianity, Islam or Buddhism have been so successful, expanding from local to worldwide religions.

But what about the existential explanation? Does it offer a more plausible account of the success of the world religions? It certainly seems to be the case. The explanation of why people around the world and across the centuries have accepted religious beliefs that they are not genetically disposed to adopt is that religion deals with their existential concern. It is a concern people have in common independently of their particular genetic set up or kinship due to their self-consciousness, and to their reflective and critical thinking. A religion like Christianity thus helps people overcome guilt and alienation, creates forgiveness and meaning. It liberates people from pure self-centredness and provides orientation and the courage to be. It provides *Homo sapiens* with a means to understand their unique form of being. Moreover, Christianity and the other world religions have done this more successfully then the tribal religions of the past, and have therefore survived on a global scale to the present time. They have had the ability to adapt successfully to changing environmental and cultural conditions.

This means also that religious beliefs are not primarily transmitted from generation to generation by genetic selection but, like moral conviction and scientific theories, through cultural selection. People teach each other not merely how to grow wheat, understand evolutionary theory or the golden rule, but also how to deal with existential concerns and to find significance in life. Consequently, it is not necessary that for a particular religion to survive its members leave more offspring than members of another religion. This is so because information about its existential superiority can be transmitted non-genetically through cultural selection or exchange. Thus people lacking genetic kinship with the original members become adherents to this new religion. If a religion is appealing for good reasons, it can then spread indefinitely through populations, jumping genetic lines and spreading much, much faster than any behavioural traits transmitted genetically.

To summarize, the Darwinian explanation has a certain plausibility dealing with tribal religion. But once a religion becomes universal, as many of the major world religions have done, this explanation will not do. Once outside the tribal context it is not plausible to maintain that religious believers who are urging us to become Christians, Muslims or Buddhists are doing so because they thereby increase their offspring in the next generation. Missionary behaviour of the magnitude present in many of the world religions provides powerful counter-evidence against a purely Darwinian explanation of religion. Religion cannot be explained as *merely* a fitness-maximizing strategy. A much more plausible explanation of these phenomena is that some religions have spread worldwide because they deal better than others with people's existential concerns and this information has spread non-genetically through cultural transmission or communication. Perhaps this is also the best scientific explanation of why religion exists and persists. Religion exists because it offers answers to questions about who we really are, why we exist and what the meaning of our life is, and what stance we should take towards experiences of death, suffering, guilt, anxiety, love and friendship.

If this is correct, religion cannot successfully be explained as merely an illusion fobbed off on us by our genes to get us to cooperate to secure our personal and inclusive fitness. Successful religion is not really 'selfish' in the sense of serving one's genes. Religion is rather 'altruistic' in the sense that it typically offers its gospel to both relatives and non-relatives, slaves and free people, male and female. So there is a clear limit to the extent evolutionary theory can explain religious belief and behaviour.

But even if this is the case, it is still possible for a scientific expansionist to maintain that science itself can be extended in such a way that it not merely answers our existential questions, but does so more successfully then any traditional religion or secular ideology. Hence, even if evolutionary theory is not able to fully explain the phenomenon of religion, it might still be possible for science to replace traditional religion. In the next chapter let us take a closer look at such an expansion of the domain of science.

CHAPTER SIX

Debunking and replacing traditional religion with science

Even if the Darwinian explanation of religion is not convincing, evolutionary biology (or science more broadly speaking) might still be of great relevance for religion in at least two different ways. It can either provide us with new information about human life and its environment that undermines (or supports) existing religious beliefs (project B) or it can perhaps even replace traditional religion and provide us with a new religion or mythology, what is often called 'scientific naturalism' (project C). In this chapter project (B) and (C) will be examined. I shall start by considering the merits of the 'debunking project'. But we cannot completely avoid taking a stand on the 'replacement project' in doing this so some conclusions about its merits will already be drawn in the first section, even though the main discussion of the replacement project will come in the subsequent section of this chapter.

Debunking traditional religion

A number of contemporary scientists seem to think that science itself is sufficient to severely undermine traditional religions such as Judaism, Christianity and Islam. William Provine, for instance, believes that 'very few truly religious evolutionary biologists remain. Most are atheists, and many have been driven there by their understanding of the evolutionary process and other science' (Provine 1988: 28). He maintains, furthermore, that

> Modern science directly implies that the world is organized strictly in accordance with mechanistic principles. There are no purposive principles whatsoever in nature. There are no gods and no designing forces that are rationally detectable. ... Modern science directly implies that there are no inherent moral or ethical laws, no absolute guiding principles for human society. ... When I die I shall rot and that is the end of me. There is no hope of life everlasting. ... Free will as it is traditionally conceived, the freedom to make uncoerced unpredictable choices among alternative possible courses of action, simply does not exist. ... There is no ultimate meaning for humans.
>
> (Provine 1988: 27–8)

The point I want to make is not that one could not maintain that there is no cosmic purpose, no God, no objective ethical principles, no immortality and no free will. The point is that Provine and other scientists claim that science *directly implies* these things. That means, I take it, that they believe that no other premises than scientific ones are needed to obtain these conclusions. What should we say about claims such as these? Can science in this straightforward way debunk traditional religion?

It seems clear that to the extent that religion tells us anything about the empirical world, its manifestations and its history, science can refute or undermine religious beliefs as well as support them. For instance, science has undermined the religious beliefs that the wind, rain and lightning are the direct manifestations of divine activity, the earth was created in six days or six thousand years and that God created an original human pair, Adam and Eve. But science has also confirmed, for instance, that Jesus of Nazareth lived and was crucified in Palestine, roughly, 2,000 years ago, that the cosmos had a beginning and that there is an underlying order in the world. Religion, however, seems to also tell us – and typically this is considered to be the important cognitive content of religion – something about a reality that underlies, interpenetrates and goes beyond the empirical world. The chief example of this, if we stick to the major theistic religions, is, of course, that God exists. Other beliefs of this kind seem to be that God created the world, loves us, created us in God's image, offers us salvation and that there is life after death and that we are here for a purpose and that, therefore, our lives have a meaning which goes beyond the one we ourselves can create.

So even if the first group of religious beliefs falls within the scope of the sciences, the latter group might not. What we should think about this issue will, however, become clearer when we have considered the concrete arguments against traditional religion offered by these scientists. But it is especially with respect to the second group of religious beliefs that we have to be careful in our examination, so that the arguments offered do not, in fact, already presuppose scientific naturalism and, therefore, beg the question. This would be the case if, for instance, the argument offered already assumed the truth of T1, namely, that the only kind of knowledge we can have is scientific knowledge.

Religion as a mechanism for survival

Let us start with some of the more general refutations of religion that scientific expansionists offer. Wilson maintains that 'traditional religious beliefs have been eroded, not so much by humiliating disproofs of their mythologies as by the growing awareness that beliefs are really enabling mechanisms for survival' (Wilson 1978: 3). We are by now familiar with this line of reasoning and also of valid objections against it. One objection is that since a religion like Christianity does not differentiate on the basis of genetic kinship, but is characterized by a missionary activity which from an evolutionary viewpoint rather help to ensure the replication of genes unlike one's own, we cannot reasonably believe that Christianity is merely a fitness-maximizing strategy. A better explanation of the phenomenon of religion is that religion helps people deal with their existential concern and that it therefore exists.

But notice that even if the existential explanation is correct (or, at least, more plausible), it can still be true that religion enables survival. It enables survival not necessarily in the sense that it promotes the survival of the key unit of selection, the genes, but in the sense that it promotes the general well-being of humans. But suppose now that it is true that religion enables survival one way or another, why should we think that this fact erodes traditional religious beliefs? Wilson does not really give us the reason why he believes this. In fact, here he seems to pick a dangerous line of argument, given his objective, because it seems as if science also enables survival.

Things happen to us that we do not anticipate and that sometimes threaten our lives and well being. We need things that are not always easy to obtain, such as nutritious food, medicine, houses, bridges and vehicles. In dealing with these things science has proved to be of great value. It enables us to control nature, and when we cannot control it, to predict it, or to adjust our behaviour to an uncooperative world. We could say that science aims to make the world *technologically* and *predictively intelligible*, and we value science because it is useful and because it helps us control, predict and alter the world. I have been trying to make a similar point about religion. We do not have to satisfy merely material needs to be alive and well. We also have to give attention to spiritual or existential needs. Our well-being thus also depends upon our ability to deal with our experiences of suffering, death, guilt or meaninglessness. In dealing with these phenomena, religion has proved to be of great value. It enables us to make sense of these existential experiences, to diagnose them, and find a way through the barriers to our well being. We might say that religion aims to make the world *existentially intelligible*.[1]

But if Wilson's line of argument is valid then both religion and science would be undermined by the fact that they enable survival. This, however, does not follow simply because the inference from the fact that 'belief *p* enables survival', to the conclusion that 'therefore one should not believe *p*' is logically invalid. If religion (or science) helps people survive, it is still possible that religion (or science) can generate truth. The only reasonable thing to say must instead be to maintain that neither scientific nor religious beliefs are undermined by the fact that they are useful for us in our predicament, that they enable survival. (It seems, in fact, more reasonable to draw the opposite conclusion. Ability to survive may be taken, at least, as an

---

[1] See Stenmark (1997) for a more detailed discussion of the different aims of science and religion.

indicator of truth.) It follows that even if Wilson is right that traditional religious beliefs have been eroded, not so much by humiliating disproofs of their mythologies as by the growing awareness that these beliefs are really enabling mechanisms for survival, this is not a good reason for rejecting these beliefs.

Wilson's argument that traditional religious beliefs ought to be rejected because they are really enabling mechanisms for survival seems reasonable only if the Darwinian explanation excludes the theological explanation (that is, that people actually have experienced a divine presence within and beyond the material world and this is the reason for being religious). If religion is *nothing but* enabling mechanisms for survival or a *wholly* material phenomenon and if it offers *purely* mundane rewards then we ought to reject the explanation religious people give of their own behaviour and it is not surprising that religious beliefs have become eroded. What I suggest is that biologists like Wilson are still not justified *as scientists* in drawing this kind of conclusion, even in a situation where there is no alternative to the Darwinian explanation (such as the existential explanation). The reason for this is that 'nothing but' reasoning is not a proper part of scientific methodology.

By using the methods of evolutionary biology we can discover that religion is at least a mechanism for survival. Suppose that this is all biologists or any other natural scientist can discover about religion (which, of course, is not surprising given the kinds of methods and theories employed), what follows? What follows is merely that any additional assertions about religion, that, for instance, it is a response to a sacred, transcendent reality, should be compatible with these biological discoveries. Evolutionary biology excludes the theological explanation only if one adds to the argument that the only things there are or, at least, that we can know anything about are things which can be investigated by the methods of evolutionary biology (or any other of the natural sciences). But premises such as T1 and T2 are, as we have seen in Chapter 2, pieces of philosophy and not pieces of evolutionary biology. Thus, scientists should resist the temptation to present their scientific account as if it were complete, realizing that such a view requires philosophical premises and not merely scientific ones.

There seems to be two strategies open to scientific expansionists at this point. They can maintain either that proper science is *a priori* committed to naturalism or that religion is a scientific hypothesis, and because of this scientists can accordingly claim that religion is a wholly material phenomenon. I shall address the first strategy in this section and the second in the following one. One response to my argument is to claim that science does not imply naturalism but rather presupposes it. Scientists, therefore, do not have to extrapolate science into an ideology or a philosophy to refute the theological explanation or to be justified in maintaining that religion is a purely material phenomenon because we cannot have any proper science without accepting one particular ideology, namely, naturalism. Naturalism is a necessary philosophical presupposition of science and therefore my objection that Wilson's claim (that science can explain religion as a wholly material phenomenon) is a non-scientific claim is not valid. That is to say, if we want science then we have to accept that science excludes the possibility of religion being something more than a wholly material phenomenon. On this point we have to choose between science and traditional religion.

This seems to be the way in which both Stephen Jay Gould and R. C. Lewontin understand the matter. Gould writes,

> I believe that the stumbling block to [the acceptance of Darwin's theory] does not lie in any scientific difficulty, but rather in the radical philosophical content of Darwin's message – in its challenge to a set of entrenched Western attitudes that we are not yet ready to abandon. First, Darwin argues that evolution has no purpose. Individuals struggle to increase the representation of their genes in future generations, and that is all. ... Second, Darwin maintained that evolution has no direction; it does not lead inevitably to higher things. Organisms become better adapted to their local environments, and that is all. The 'degeneracy' of a parasite is as perfect as the gait of a gazelle. Third, Darwin applied a consistent philosophy of materialism to his interpretation of nature. Matter is the ground of all existence; mind, spirit and God as well, are just words that express the wondrous results of neuronal complexity.
>
> (Gould 1977: 12–13)

Lewontin claims that scientists have

> a commitment to materialism. It is not that the methods and institutions of science somehow compel us to accept a material explanation of the phenomenal world, but, on the contrary, that we are forced by our *a priori* adherence to material causes to create an apparatus of investigation and a set of concepts that produce material explanations, no matter how counterintuitive, no matter how mystifying to the uninitiated. Moreover, that materialism is absolute, for we cannot allow a Divine Foot in the door. The eminent Kant scholar Lewis Beck used to say that anyone who could believe in God could believe in anything.
>
> (Lewontin 1997: 31)

Gould's and Lewontin's view appears to be that scientists are *a priori* materialists or naturalists. Gould maintains that if one accepts evolution, one must first embrace naturalism since evolution is inseparable from naturalism. (One must also accept certain implications of evolution and materialism, namely that there is no purpose to life and the universe, and that mind, spirit and God are merely words whose referent apparently can be completely reduced to neuronal activities, but more on this latter.) Lewontin thinks this is not merely true about evolution, but of all science. Science is inseparable from naturalism. Consequently, Wilson does nothing wrong as a scientist when he concludes that religion is nothing but a set of enabling mechanisms for survival, a wholly material phenomenon, and that it offers purely mundane rewards because that is merely to apply a consistent philosophy of naturalism, an essential part of the scientific enterprise.

But serious objections can be raised against the idea that proper science is *a priori* committed to naturalism. For example, Isaac Newton, Michael Faraday and Arthur Stanley Eddington, to name a few, were theists and not naturalists. Were they, therefore, not good scientists? Perhaps Gould and Lewontin would agree that they were good scientists, but maintain that they were not really consistent in their scientific methodology. If they had been, then they would have realized that scientists are and must be naturalists. I have doubts whether that reply would hold up under closer scrutiny, but a more fundamental objection is that it appears that, for instance, Newton's laws are true whether naturalism is true or false, that is, whether it is the case that matter or physical nature alone is real. The same is true about evolutionary theory. It can be the case that life has progressed from simple to complex forms by means of natural selection, whether or not it is true that reality consists of nothing but matter in motion.

However, it is true that science presupposes some kind of naturalism. To that extent Gould and Lewontin are right, but they fail to see that there are different forms of naturalism and that science does not presuppose a kind of naturalism that entails that traditional religions must be explained as purely material phenomena. Gould's and Lewontin's mistake is that they overlook a crucial distinction within scientific methodology, in that they fail to acknowledge the difference between metaphysical and methodological naturalism. *Metaphysical naturalism* maintains, roughly, that matter or physical nature alone is real and that everything that exists (life, mind, morality, religion and so on) can be completely explained in terms of matter or physical nature. These claims, however, are not scientific theories or assertions. They do not belong to biology, chemistry, physics or psychology. *Methodological naturalism*, on the other hand, is, roughly, the view that scientific explanations are to be in terms of only natural entities and processes. It says that scientists should seek naturalistic explanations of all natural events.[2] Methodological naturalism lays down which sort of study qualifies as scientific. Naturalists, Christians, Buddhists and Marxists alike must in pursuing science be satisfied with this kind of explanation. As scientists they must be content with not giving ultimate explanations of reality.

Methodological naturalism does not entail metaphysical naturalism. One

---

[2] See also Clayton (1997: 169–77) for an instructive discussion of naturalism in scientific practice.

can accept the former but reject the latter without any logical inconsistency. Furthermore, methodological naturalism is the only form of naturalism science needs to work. That is to say, it is sufficient that only natural entities and processes will enter into the theories and explanations given by science for this enterprise to be successful; no further commitments about whether or not there exist objects and properties other than natural ones is required. Therefore, science is not *a priori* committed to a form of naturalism which excludes religion and which refutes the theological explanation.

Let us see whether the second strategy that scientific expansionists can use to exclude a theological explanation of religious belief and behaviour and justify them in taking their explanation to be complete is more successful. Even if one admits that scientists should resist the temptation to present scientific explanations as if they were complete and that science is not *a priori* committed to metaphysical naturalism, one can still maintain that science excludes the truth of traditional religion (in the sense that it is false that it is an encounter with a sacred, transcendent reality) because religion is a scientific hypothesis and, therefore, a rival to evolutionary theory and as such has been refuted by scientific discoveries.

Religion as a scientific hypothesis

Thus one more general claim we have to consider (besides the idea that the survival function of traditional religion undermines its credibility) before looking at some of the more detailed arguments is the idea that religion is a scientific or quasi-scientific hypothesis or theory. If this is correct, it can be shown within a scientific discourse that evolutionary theory is a much better supported theory than traditional religion and that we therefore ought to reject the latter.

Dawkins, like Wilson, thinks that science and religion compete, not merely because (a) science can be expanded to deal with our existential concern but because he believes that (b) traditional religions actually are scientific or quasi-scientific hypotheses or theories:

> I pay religions the compliment of regarding them as scientific theories and ... I see God as a competing explanation for facts about the universe and life. This is certainly how God has been seen by most theologians of past centuries and by most ordinary religious people today. ... Either admit that God is a scientific hypothesis and let him submit to the same judgement as any other scientific hypothesis. Or admit that his status is no higher than that of fairies and river sprites.
>
> (Dawkins 1995b: 46–7)

Either religious believers have to treat belief in God as a scientific hypothesis or they have to admit that it has merely a fairy-tale status, it is a superstition. These are the only options that Dawkins apparently thinks are possible.

Even though Wilson does not explicitly state that he sees traditional religion as a scientific hypothesis, he, at least, sometimes writes as if this would be his case. Wilson maintains, for instance, that

> the reasons why I consider the scientific ethos superior to religion [are]: its repeated triumphs in explaining and controlling the physical world; its self-correcting nature open to all competent to devise and conduct the tests; its readiness to examine all subjects sacred and profane; and now the possibility of explaining traditional religion by the mechanistic models of evolutionary biology.
>
> (Wilson 1978: 201)

If religion is a scientific hypothesis then it competes with other scientific hypotheses and a possible outcome of such competition is, of course, the refutation of the religious hypothesis. Is this a reasonable understanding of traditional religion? To be able to answer that question we have to address two issues: (1) is religion a hypothesis? and (2) is religion a scientific or quasi-scientific hypothesis? These are two separate issues because it is possible that something can be a hypothesis, but nevertheless not be a scientific hypothesis. For instance, my belief that you are tired and perhaps did not slept much last night is a hypothesis I reach by looking at your face, but it is not a scientific hypothesis.

To be able to answer these questions, however, we have to narrow our focus. Religion comes in many radically different versions, so it is difficult to determine the answer to these two questions in respect of all that variety. Let us, therefore, merely focus on belief in God as it is traditionally understood by Jews, Christians or Muslims (that is, theism), which after all seems to be the main target of both Dawkins and Wilson. (In what follows the concept 'traditional religion' will primarily refer to these three major theistic world religions.)

We first need to know what a hypothesis is. I suggest that by a hypothesis we mean *an assumption made in order to explain a phenomenon or an event*. For instance, when I see footprints outside my kitchen window I explain this by assuming that a human being has passed by my window. By speculation this hypothesis could be made more precise. The size of the footprints together with their depth seems to indicate that a large human being, probably a man, has passed by my window. All hypotheses have in common the feature of speaking about things beyond the evidence. They tell us more than we can see for ourselves. My hypothesis tells me that a male person passed by outside my window, even though all I can see are footprints in the mud. Thus, what makes something a hypothesis is that it explains or helps us understand a phenomenon. However, different hypotheses differ in their credibility. Some of them are merely 'working' hypotheses, which are only tentatively accepted. They are believed to be true but need to be investigated further. They are, perhaps, only minimally tested. Other hypotheses are believed not only to be true but in addition one thinks that there is no need to investigate them further. These are hypotheses with a high degree of certitude. Hypotheses of this latter kind we take for granted in life in general, in politics and in science and we feel no need to question them. For instance, evolutionary theory is a hypothesis (or rather a set of hypotheses) in this sense because it functions as a paradigm for much of the research done in contemporary biology.

Is belief in God a hypothesis? There are reasons to doubt this because for religious people God is typically taken to be an experienced reality and not a derived entity of some sort. Their faith is not a hypothesis invented to explain natural events in the world. It is rather an expression of an experience of something holy, something beyond the mundane and which is of supreme value and, therefore, worthy of worship and prayer. In short, they have had an encounter with what they experience to be a divine reality. This is well captured by Keith Ward when he writes:

> Many people, perhaps most, occasionally experience a sense of something transcendent, something beyond decay and imperfection ... Perhaps religious faith begins, for many of us, in such small epiphanies, in 'a sense and taste for the Infinite.' It is from such glimpses of a spiritual reality underlying this phenomenal world that one may develop a desire to seek a deeper awareness of it, and, if possible, seek to mediate its reality in the world. If that happens, religious faith is born. Worship and prayer are, basically, ways of deepening this awareness and transforming the self to reflect and mediate the divine spirit.
>
> (Ward 1996: 102–3)

The primary aim of a traditional religion such as Christianity is, thus, not to explain and predict observable events, but to transform people's lives as a response to an encounter with a divine reality. A reality that believers claim helps them to deal with experiences of suffering and anxiety, and which gives their lives a meaning. It involves one's deepest personal commitments. Belief in God is, therefore, probably more closely related to belief in other persons than belief in the existence of genes, electrons, planets or any other scientific stuff. This explains why it could be deeply problematic to treat belief in God as a tentative hypothesis. To understand, for instance, one's relationship with the beloved, in love affairs, as a tentative hypothesis (as something one should test by actively searching for counter-evidence) would seem to destroy the very foundation necessary for a loving relationship – it undermines the trust and loyalty that must exist between two lovers. But this seems also true about belief in God because it is not just a matter of mentally assenting to a set of propositions, but of trusting God (see Stenmark 1999: 140). It is, therefore, probably correct to say, as Ward does, that to treat belief in God as a tentative hypothesis would typically be 'like saying that a good marriage is best achieved by always seeking evidence of infidelity' (Ward 1996: 97).

What follows? For one thing, it follows that it is not surprising if believers or theologians do not consider their religious faith to be a hypothesis which is either scientific or commonplace. Consequently, to discuss the rationality of religious commitment as if people's belief in God *is* for them a hypothesis, is, therefore, to seriously misunderstand the nature of their religious faith. Moreover, to treat the issue as if these people's belief in God *ought to be* a hypothesis is unjustified. The reason why is perhaps best explained by going back to the everyday situation I described above. Imagine that this morning I was in my kitchen and saw a man pass by outside my kitchen window. Assume now that a sceptic would claim that it is rational for me to believe this only if I treat my belief as a hypothesis. In other words, I must first go out and check whether there are any traces left of the man and, second, for me to be rational in accepting this conviction I have to believe this on the basis of this evidence. Suppose further, that neither the sceptic nor I are able to find any traces, any mark of footprints. The sceptic's conclusion is clear. He claims that it is irrational for me to hold this belief. Stammering I maintain, however, 'But, but, but I saw him!' I would claim that many religious believers are in an analogous situation in respect to their belief in God.[3] They have experienced God's presence in their lives; they have encountered at least glimpses of a divine reality. In a similar way that I believe that a man passed by my window they believe that God exists and believe in God's presence in their lives. If this is correct, it is unreasonable to demand that these people should view their belief in God as a hypothesis of some kind. Their belief, just like mine, is directly experientially grounded. This in contrast to the belief the sceptic would have obtained if there had been footprints outside my window. The sceptic's beliefs would then have been indirectly experientially grounded because he or she would have derived the existence of this man from certain facts or evidence (such as the presents of footprints, their depth, etc.). The same is true in all other cases when hypotheses are assumed or proposed.

Notice, however, that the things we believe which are directly experientially grounded can nevertheless be something we are uncertain about. I was in my situation quite certain because I clearly saw a man outside the kitchen window. But the situation could just as well have been that I merely saw the contours of a man in the window, but because my focus was elsewhere or because the sun shone through the window I am uncertain about what I really saw. The same thing can be true about religious belief. Some religious believers have experienced merely glimpses of what they think is God in their lives and, therefore, believe in God but still feel uncertain. Others have had more profound experiences of God's presence, experiences which are harder for them to doubt. Augustine writes, for instance, 'far off, I heard your voice saying *I am the God who IS* ... and at once I had no cause to doubt.

---

[3] See Alston (1991) for an attempt to justify this analogue.

I might more easily have doubted that I was alive than that Truth had being' (Saint Augustine, *Confessions* VII: 10).

Thus, religious believers do not have to treat their belief in God as a hypothesis because for many of them this belief is directly experientially grounded. But would this mean that no religious believer whatsoever or any other person (say Dawkins or Wilson) would be justified in seeing belief in God as a hypothesis? I do not think this follows. Consider an example from another context. My belief that my wife loves me is for me not a hypothesis. However, this does not mean that it could not be a hypothesis for somebody else. Suppose, for example, two persons hear me say that my wife loves me. One of them believes me but the other does not. So they decide to treat my belief as a hypothesis and try to collect evidence for or against its truth that they can both accept. I can see nothing problematic about that. The same seems to be true about religion. Take a person who has never experienced God's presence at first hand. Suppose she attaches some importance to the believers' testimony. But what makes her believe in God is after all that she thinks that she can see God's footprints in the universe. The existence and beauty of the cosmos convince her that God exists and is worthy of worship. This person's belief in God is, thus, indirectly experientially grounded. Perhaps from a religious perspective there is a better way of obtaining a belief in God, but there is no reason to doubt that this is a genuine faith. This person has hardly misunderstood belief in God in some fundamental way. Or take sceptics who wonder about the rationality of religious faith. They know about the believers' testimony but that is not sufficient for them. Instead they wonder whether it would not be reasonable to assume that if the God who believers are talking about really exists, really has created heaven and earth, then somehow this divine reality should have left some marks, some footprints in the world. The non-believer might even, like J. J. C. Smart, want to be a believer, if only there were sufficient evidence of this sort supporting belief in God (Smart 1996: 6).

Hence we should not confuse the questions, '*Must* or *should* believers see their belief in God as a hypothesis?' and '*Can* a believer (or somebody else) see belief in God as a hypothesis?' The answer to the first, as we have seen, is that religious believers need not treat their belief in God as a hypothesis because for many of them this conviction is directly experientially grounded. The answer to the second is that religious believers should understand their belief in God as a hypothesis if it is indirectly experientially grounded and the same is true about sceptics or 'seekers'.

Where does this leave us? One thing we realize is that there is a crucial difference between religion and science on this point. Scientists, whether or not they believe a particular theory to be true, always treat it as a hypothesis in the sense that it is an assumption made to explain a phenomenon and it needs to be supported by other things which function as evidence, whereas religious believers and sceptics do not share such an agreement when debating whether or not it is true or rational to believe in God. That is to say, scientists who disagree face a situation analogous to one in which two persons are looking at some marks in the ground that look like two footprints, and person *A* believes that a human being has been standing there, whereas person *B* does not think so. (They interpret the evidence in different ways.) The religious believer and the sceptic, on the other hand, typically face a situation analogous to one in which person *A* claims that she has seen a man passing by outside the window and person *B* doubts this to be the case (perhaps on the grounds that person *A* was drunk at the time or that no footprints can be found outside the window).

Second, and of great importance for our inquiry, even if we cannot find any evidence of the presence of God in the physical world, this would not automatically undermine the credibility of religious faith, whereas the credibility of a proposed scientific theory would typically be undermined if no evidence could be found to support it. The reason for this is that belief in God is not held on the basis of other beliefs that function as evidence; instead it is directly experientially grounded. Just as my belief that I have seen a man pass by outside my window is not automatically undermined by the lack of footprints in the grass, people's belief in God is not necessarily challenged if God has not left any physical footprints detectable by the sciences. But that is, of course, not to deny that if such divine footprints could be found, it would strengthen the case that God really exists, just as the credibility of my belief in this man passing by outside my window would be strengthened if footprints could be found in the grass. The discrepancy, expressed in the terminology developed in Chapter 2, is that belief in God is typically a direct knowledge claim (it is more than this of course), whereas belief in electrons, natural selection, mutations and so on are indirect knowledge claims.

Third, the beliefs within a person's noetic structure (that is, the cluster of accepted beliefs and their epistemological relations to one another) can function in different ways in different contexts and it would, therefore, not necessarily be improper for religious believers to sometimes treat their religious faith as a hypothesis (or set of hypotheses).[4] I would typically not treat my belief that my wife loves me as a hypothesis, but in a context where there are flowers on the table of my office and I cannot find any card telling me who put them there, one possible explanation might be that my wife left them there as a sign of her love and affection, and the reason why there is no card can be explained by assuming that it fell off on the way to my room but my wife did not notice this. In this particular case my belief that my wife loves me functions as a hypothesis explaining a particular phenomenon, the flowers on my desk. I suggest that religious believers can in a similar way use

---

[4] See Plantinga (1983: 48–50) for a more detailed discussion about what a noetic structure is.

their belief in God as a hypothesis explaining a certain range of phenomena, for instance consciousness and the order and beauty of the world; and this has also been the case.

Thus belief in God can, at least sometimes, function as a hypothesis for the believer. But is it in these circumstances a *scientific* hypothesis? Dawkins maintains, as we have seen, that God is a explanation for facts about the universe and life, and therefore is a scientific theory.[5] We have seen that this is misleading. It is misleading because belief in God is not a theory or hypothesis invented to explain particular facts about the physical and biological world like a scientific theory is, but is rather taken by religious believers to be the outcome of an encounter with a divine reality, a reality that these believers claim helps them to deal with experiences of suffering and anxiety, and which gives their lives a meaning. Dawkins' claim also gives the impression that if religion cannot successfully compete with science then religion is superfluous and undermined by science. But this is to miss the point of religion. It is to make possible a relationship with a divine Other, a relationship deepening through worship and prayer; it does not seek to offer a competing explanation to scientific theories for facts about the universe and life.

Religious theories of design

It is still possible that belief in God can *in certain circumstances* function as a hypothesis or theory and I think Dawkins is right that it has functioned within the sciences as an explanation of the origin of life. Almost everybody up to the second half of the nineteenth century accepted what Dawkins call the 'Conscious Designer theory' (Dawkins 1986: 4). The conscious designer theory is, roughly, the idea that a cosmic intelligent designer created the biological species that exist in a way analogical to the way we create artifacts. This meant that species were taken to be separately created by this cosmic intelligent designer (that is, by God) in such a way that they were not genealogically related and did not evolve from common ancestors. This was the theory available and which was defended by scientists of that time in one version or another. There simply was no other theory of the origin of life available to scientists. Thus, belief in God functioned as a scientific explanation within biology (see Lindberg and Numbers (1986) and Brooke (1991)). In fact, the apparent design scientists found in nature provided strong evidence for the existence of God (or of a divine designer). Dawkins agrees with this, and he tells us that he 'could not imagine being an atheist at any time before 1859, when Darwin's Origin of Species was published' (Dawkins 1986: 5). Many philosophers have thought that David Hume in his criticism

---

[5] In this study I make no distinction between what a theory is and what a hypothesis is.

of Paley's design argument has established the plausibility of the atheistic position. Dawkins disputes this,

> what Hume did was criticize the logic of using apparent design in nature as *positive* evidence for the existence of God. He did not offer any *alternative* explanation for apparent design, but left the question open. An atheist before Darwin could have said, following Hume: 'I have no explanation for complex biological design. All I know is that God isn't a good explanation, so we must wait and hope that somebody comes up with a better one.' I can't help feeling that such a position, though logically sound, would have left one feeling pretty unsatisfied, and that although atheism might have been *logically* tenable before Darwin, Darwin made it possible to be an intellectually fulfilled atheist.
>
> (Dawkins 1986: 6)

Dawkins clearly has a point here because philosophers, in contrast to scientists, have sometimes neglected the importance of having an alternative theory before rejecting a previously accepted theory. But not until the publication of *Origin of Species* was such a rival theory available to atheists (or naturalists, as we have called them). From there on, but not before, was it possible to be an 'intellectually fulfilled' naturalist.[6] This means, however, that for naturalists evolutionary theory is the only game in town. For someone who believes that matter alone is real and that there is no God, evolution in some form or other is the only possible answer to the question of the origin of life. This is well worth stressing because it means, as Ernan McMullin points out, that the fortunes of atheism as a form of intellectual belief would depend upon the fortunes of the theory of evolution (McMullin 1991: 58). It is also worth noting that, at least in some places, Dawkins writes as if he would stick to Darwin's theory no matter what. For instance, 'even if there were no actual evidence in favour of the Darwinian theory (there is, of course) we should still be justified in preferring it over all rival theories' (Dawkins 1986: 287). The theory's profound importance for naturalism as a world view might explain a statement such this.

Is there a symmetry here, so that it is also true that the conscious designer theory is the only game in town for religious believers? To be able to answer this question we have to know exactly which religious beliefs are undermined by evolutionary theory. Evolutionary theory tells us that life on earth came into existence through the gradual process of evolution from a primitive soup of matter by means of heritable variation in fitness. This means roughly that scientists do not need anything more than the laws of physics, the primitive soup of matter, natural selection, genetic mutations, self-replication and a long period of time to explain the origin and variation of life. Therefore, Dawkins writes, we now know that

---

6 This has been stressed by Plantinga (1991) and McMullin (1991).

> All appearances to the contrary, the only watchmaker in nature is the blind forces of physics, albeit deployed in a very special way. A true watchmaker has foresight: he designs his cogs and springs, and plans their interconnections, with a future purpose in his mind's eye. Natural selection, the blind, unconscious, automatic process which Darwin discovered, and which we now know is the explanation for the existence and apparently purposeful form of all life, has no purpose in mind. It has no mind and no mind's eye. It does not plan for the future. It has no vision, no foresight, no sight at all. If it can be said to play the role of watchmaker in nature, it is the *blind* watchmaker.
>
> (Dawkins 1986: 5)

Because evolutionary theory turned out to be better supported by the evidence than the conscious designer theory, it replaces the latter within the biological sciences. This means that *evolutionary theory undermines the religious belief that God has created the different species analogously to how we create artifacts*. As Dawkins points out, Paley's 'analogy between telescope and eye, between watch and living organism, is false' (Dawkins 1986: 5). So if there is a God, it is not true that this God created humans and other complex organisms in a similar way as to how a human designer makes a watch.

If, moreover, the conscious designer theory (or let us instead call it the 'artifact designer theory' because what is essential to the theory is the analogy between artifacts and cases of biological complexity) is the only game in town for religious believers, its replacement within the sciences has tremendous implications for the plausibility of traditional religion. Science once supported traditional religion, but now it supports atheism or naturalism. It is, thus, not possible any longer to be an intellectually fulfilled religious believer. This conclusion, however, is plausible only if evolutionary theory undermines not merely the religious belief, 'God has created the different species analogously to how we create artifacts', but also the more basic religious belief that 'God has created the world'. But is it not possible that if there is a God, this God could have accomplished the creation of this world in a number of different ways? It seems possible that such a God could, for instance, have chosen to create the world by means of the process of evolution. God could have set up the initial condition, that is, the primitive soup of matter and the laws of nature, and these conditions eventually gave rise to complex organisms, among them human beings. God on this account, let us call it the 'origin designer theory', would use natural laws and natural selection in order to bring about complex forms of life, especially conscious beings.

This possibility[7] suggests that the conclusion that Dawkins draws from the fact that evolutionary theory can explain the development of life is not right. Dawkins maintains,

---

[7] See Ward (1996: 77f.) for an interesting attempt to develop an alternative way of relating evolutionary theory and theism.

> The claim of the existence of God is a purely scientific one. Either it is
> true or it is not. A universe with God would be completely different from
> one without ... If you're deeply steeped in evolution, you see that it is a
> way to get complex designs out of nothing. You don't need God.
>
> (Dawkins 1992: 3)[8]

He also writes that his book *The Blind Watchmaker*, is 'written in the
conviction that our own existence once presented the greatest of all mysteries,
but that it is a mystery no longer because it is solved. Darwin and Wallace
solved it ...' (Dawkins 1986: ix). But this conclusion that one can get complex
designs out of nothing and that therefore our own existence is no longer a
mystery, does not follow from evolutionary theory. This is so because
although the Darwinian explanation is a *correct* explanation, it is not, as
Swinburne points out, an *ultimate* explanation of our existence:

> For an ultimate explanation we need an explanation at the highest level of
> why those laws rather than any other ones operated. The laws of evolution
> are no doubt consequences of laws of chemistry governing the organic
> matter of which animals are made. And the laws of chemistry hold
> because fundamental laws of physics hold. But why just those
> fundamental laws of physics rather than any others? If the laws of physics
> did not have the consequence that some chemical arrangement would give
> rise to life, or that there would be random variations by offspring from
> characteristics of parents, and so on, there would be no evolution by
> natural selection. So, even given that there are laws of nature (i.e. that
> material objects have the same powers and liabilities as each other), why
> just those laws?
>
> (Swinburne 1996: 60)

But since evolutionary theory does not provide us with an answer to this kind
of question, the mystery of our existence is not yet solved and evolution is not
a way to get complex designs out of nothing. Both naturalism and theism, on
the other hand, provide answers to this question. Therefore, theism does not
compete with science, but it does compete with naturalism. Naturalists
maintain that it is merely an accident that these laws happen to operate and
that the primitive soup of matter had the particular constitution it had. The
best ultimate explanation of the constitution and general order of nature is that
it is a work of pure chance. Theists disagree and claim instead that the world
is God's creation and it is, therefore, not merely an accident that the laws of
nature happen to operate and the primitive soup of matter had the particular
constitution it had. Some of them maintain that this is even what we could

---

8 Dawkins' argument seems to be: (1) if a claim is either true or false then it is a scientific claim;
   (2) the claim that God exists is a claim that is either true or false (probably false); (3) therefore,
   it is a scientific claim. This is true, however, only if we presuppose that T1 is true. But we have
   seen, in Chapter 2, we have good reasons to reject T1; in fact, it is even self-refuting.

expect if there is a God. Hence, they would agree with Dawkins that a universe with God would be different from one without. If there is a God and this God intended to create a world containing persons who could have experiences, thoughts, make choices, develop virtues and so on, then we would expect the world to be ordered in such a way that such beings could flourish. So if God exists we could expect that material objects would not behave totally erratically, but rather expect, as Swinburne points out, that medium-sized material objects would behave in a regular way and therefore be detectable by humans (Swinburne 1996: 60f.). Stones would be solid objects and fall to the ground and not have these propensities one day and the other fall apart and float in the air and so on. But such an orderly world is nothing we would necessarily expect if naturalism is true. On this account it is just a marvelous coincidence that we can understand what is going on in the world, that all material objects have the same simple powers and propensities as each other and, thus, that matter is constituted in such a way that life could evolve.

Two conclusions, at least, follow from our discussion. First, it is not the case that *evolutionary theory undermines the religious belief that God has created the world or the universe*. Second, whereas the fortunes of naturalism as a form of intellectual belief depend upon the fortunes of the theory of evolution, the future of traditional religion does not depend upon the fortunes of the artifact designer theory because the belief that God has created the world is compatible with evolutionary theory.

A number of philosophers, scientists and theologians even think that recent development in science has provided us with evidence which makes the ultimate theistic explanation of life more likely than the ultimate naturalistic explanation.[9] Scientists have discovered that even a marginal difference in the initial conditions of the universe would have ensured that no life ever evolved anywhere. For life to emerge at all, the rate of expansion of the universe, the force of gravity, the weak and strong nuclear forces and innumerable other physical and cosmic conditions had to fall within a very narrow range. If not, the universe could never have produced hydrogen atoms, supernovae, carbon, water and other elements essential to life. The universe seems to be 'fine-tuned' to support intelligent life. The very success of science in showing us how deeply orderly the natural world is provides grounds for believing that there is an even deeper cause of that order.

We have here a new design argument but one which is quite different from Paley's. It is not an argument from analogy but an inference to the best explanation; it takes as premises facts from astronomy, chemistry and physics rather than the facts of the biological sciences; and the examples of design or fine-tuning are the initial physical conditions and the laws of nature rather

---

[9] See, for instance, Murphy (1993), Polkinghorne (1998), Swinburne (1996) and Ward (1996).

than the particular outcome of such laws, as for instance, the eye. So perhaps God has left some footprints in the physical universe after all, at least, it is still possible to believe that the world is God's creation. The acceptance of evolutionary theory does not rule out the possibility of being an intellectually fulfilled religious believer.

These divine footprints were not to be found where Paley and many others thought they would be to be found, namely in specific cases of biological complexity, but in the general order of nature and the fine-tuning of the initial physical conditions. This means, however, that the origin designer theory is a hypothesis, but it is not like the artifact designer theory a scientific hypothesis. This is so because the origin designer theory attempts to explain what science takes for granted, namely, that all material objects have the same simple powers and propensities, and that matter is constituted in such a way that life could evolve. These things constitute the framework of science itself because, as Roger Trigg points out, 'For science to be possible, the world has to be ordered and structured in some way' (Trigg 1993: 193). It is this order and structure of the world that the origin designer theory attempts to explain. It is a theory which is introduced *to explain why science can explain things*. By assuming the existence of God we can explain why it is not surprising that the physical world is such that it can be understood by us. Why it and not merely we embody rationality. Hence, this version of a divine designer theory is a *metascientific* or *philosophical* hypothesis and not a scientific one. Just like its rivals, it is engaged not to explain things within the scientific domain but to explain phenomena such as the very existence, orderliness and evolutionary potential of the universe, which constitute the framework of science itself. An alternative theory is, of course, the naturalistic theory, which maintains that it is completely accidental that the universe had such natural laws and such a primitive soup of matter from which life could evolve, even when a marginal difference in those initial conditions would have ensured that no life ever emerged anywhere.

I personally believe it is more likely that the theistic explanation is true than the naturalistic explanation, but that is beside the point. The point is that when we participate in this kind of discussion we have left the sciences behind and entered the arena of philosophy. When we argue for or against these two ultimate explanations of reality, we are not doing science anymore. This does not mean, of course, that scientific information is irrelevant to the discussion, it only means that the explanations offered are not scientific ones. Science can and should be agnostic about this debate because scientific theory construction and justification will work whether or not it is true that God exists or that matter is what is ultimately real. Moreover, we have seen that it is not correct to assume, as Dawkins does, that religion is undermined if evolutionary theory is true, although he is right that naturalism is seriously undermined if evolutionary theory (or something in the neighbourhood) turns out not to be true. Evolutionary theory does not show that belief in God is superfluous because it is not a hypothesis invented primarily to explain the events in the world, but the outcome of an encounter with what believers think is a divine reality. What is true is that evolutionary theory undermines one conception of divine creation (the artifact designer theory) but as long as other conceptions are possible (such as the origin designer theory), belief in God and the central theistic belief that God created heaven and the earth are not undermined by the discoveries of modern biology.

But perhaps this conclusion is premature. There are indications in the writings of scientific expansionists that they think, first, that evolutionary theory alone (or in combination with other scientific theories) implies a non-purposive or meaningless universe and, thus, also a godless one and, second, that the theory of natural selection indicates that the evolution of life is a ruthless struggle for survival, something which is incompatible with belief in God. Let us, therefore, take a closer look at the purported scientific arguments offered for these conclusions.

Evolutionary theory and the problem of evil

Evolutionary biology gives us new insight into the enormous amount of suffering present in the natural world. It seems as if the process of evolution is insensitive to the pain of living beings and it, therefore, appears questionable whether it is reasonable to believe that a loving and caring God could be the creator of that system or the initial conditions that made such a system possible. Dawkins writes,

> If Nature were kind, she would at least make the minor concession of anesthetizing caterpillars before they are eaten alive from within. But Nature is neither kind nor unkind. She is neither against suffering nor for it. Nature is not interested one way or the other in suffering, unless it affects the survival of DNA.
>
> (Dawkins 1995a: 131)

He therefore concludes,

> The universe we observe has precisely the properties we should expect if there is, at bottom, no design, no purpose, no evil and no good, nothing but blind, pitiless indifference. ... DNA neither knows nor cares. DNA just is. And we dance to its music.
>
> (Dawkins 1995a: 133)

The idea is that biological discoveries provide us with new evidence that is not compatible with theism or, at least, makes naturalism more likely to be true than theism.

It comes, of course, as no surprise for theists that Nature is not interested one way or another in suffering (whether or not it affects the survival of DNA) and is neither kind nor unkind because they (and typically also naturalists) do not consider Nature to be an entity that can be interested or uninterested, kind or unkind. For something to have interests or to be kind or unkind, the entity in question must at least have consciousness; and persons would be the paradigm example of such entities. But Nature is just not that kind of entity and it therefore makes no sense to talk about Nature the way Dawkins does. So that cannot be the issue. What he probably means is that God, who is assumed to be the creator of nature, is believed to be kind and should, therefore, be interested one way or another in suffering and even care about living things. But evolutionary biology cannot find any traces of such a kindness; rather it finds traces of unkindness and meaningless suffering in nature. Therefore, evolutionary biology undermines religious belief. More exactly, *evolutionary theory undermines the religious belief that a powerful and benevolent God could have created complex organisms by means of evolution.* But Dawkins assumes that evolutionary biology does more than that; in fact it shows that the world is exactly like we would expect it to be if naturalism is true. Thus, evolutionary theory not merely undermines traditional religion but confirms naturalism.

Dawkins raises here a serious challenge to religious faith because the idea that God is good, even perfectly good (as well as tremendously powerful), is a central conviction in many religions, and especially so in Christian faith. Hence, if he is right then there really is a serious clash between science and religion (or rather religions similar to Christianity) and naturalism might turn out to be a very promising alternative.

Let us try to state the premises of Dawkins' argument more clearly, before we evaluate it. The argument seems to be:

(1) Evolutionary biologists have discovered instances of evil, suffering or unkindness in the natural world of a magnitude previously unknown.
(2) A good and powerful God would not allow evil, suffering or unkindness of such a magnitude as found in the natural world.
(3) Therefore, it is unlikely that a good and powerful God exists.

Furthermore,

(4) Such an amount of evil, suffering or unkindness in the natural world is what we should expect given naturalism.
(5) Therefore, naturalism (everything else equal) is more likely to be true than theism.

An argument based on the existence of evil in the world is the strongest kind of argument I know of against belief in God (as it is understood by Christians and many others), but despite this the question that is crucial for our inquiry is whether this argument can be a *scientific* argument. This is important because Dawkins gives the impression that evolutionary biology itself shows that belief in God is misplaced. We do not need anything more than evolutionary theory to undermine traditional religion (because they are after all both scientific theories). Let us, therefore, examine premise (1) in light of this question: *Can evolutionary biologists* qua *biologists really discover evil, suffering, unkindness or meaninglessness (or add any other of the natural sciences if you like)? Are evil, suffering, unkindness or meaninglessness properties that biologists can discover by using merely their own methods and instruments and experiments*? Of course, sometimes it sounds like biologists are discovering moral (or immoral) properties all the time. We can read that gorillas and wrens 'lie to one another' and get 'cheated'. Hyenas are truly 'murderous' and there is 'warfare' and 'slavery' amongst ants. Duck can commit 'rape' and some bird males are said to be 'sneaky fuckers'. Organisms and even genes are said to be 'selfish'. Dawkins tells us that the main argument of *The Selfish Gene* is that we and all other animals are 'like successful Chicago gangsters' and that his 'purpose is to examine the biology of selfishness and altruism' (Dawkins 1989: 1–2).

But remember that what Dawkins is actually talking about is when the outcome of animals' behaviour is such that it decreases/increases other animals' chances of survival and reproduction in accordance with the advantage/disadvantage to their own reproductive fitness. So when a bear kills a moose or a fox kills a rabbit we have examples of this kind of behaviour. Or more fundamentally, Dawkins focuses on what he takes to be the key unit of natural selection, the genes. When a bear kills a moose or a fox kills a rabbit, it is ultimately the genes that cause this behaviour, and what counts biologically speaking is whether the chances of survival of these animals' genes are improved by this behaviour. But whether this behaviour is morally good or evil falls outside the scope of biology (or any other of the natural sciences). Evil is not a property that can be discovered by using the methods of biology. It is rather something that requires a moral sensibility and, therefore, only moral agents can discover evil states of affairs. (The point is, of course, not that biologists lack a moral sensibility but that this is something that they possess not *as* biologists but *as* moral agents.)

Dawkins' Swedish colleague Torbjörn Fagerström makes this limitation of biology clear when he writes,

> The theory of evolution does not contain any independent evaluation about what are good or bad features. Therefore there are no independent scales which one can use to estimate the adaptive value that a certain feature has; this value can only be measured on the scale that is given by the actual environment. Suppose that there is a disease that reduces pheasants' ability to escape the goshawk. Goshawks will then mainly capture those pheasants that have the disease and we can observe that the disease has negative adaptive value for pheasants *in an environment*

> *where we can find goshawks.* If one, on the other hand, eliminates all goshawks, then the disease is not any longer a handicap and the number of sick pheasants will therefore increase in a goshawk free environment. But these are neither better nor worse pheasants than the healthy ones; they are merely pheasants that are adapted to live in an environment free from goshawks in the same way as healthy pheasants are adapted to live in an environment where there are goshawks. ... Darwinism does not provide us with values about whether [a particular state of affairs] is a better or a worse state of affairs. Period!
>
> (Fagerström 1994: 44–5, my translation)

The fact that some biologists write *as if* animals and genes were ethical agents will not change a thing in this regard, even if it, of course, obscures the issue tremendously. But the things biologists like to say can be said in descriptive language without loading the biological discourse with moralistic and pejorative overtones. For instance, Dawkins' thesis is that 'a predominant quality to be expected in a successful gene is ruthless selfishness' and he, therefore, talks about 'the gene's law of universal ruthless selfishness' (Dawkins 1989: 2–3), but the scientific claim (whether true or false) in this case is nothing more than that the genes cause the organism to react in such a way that the survival of the genes is secured. If, however, Dawkins' theory is merely a claim about causes and effects then it does not reveal anything about values in nature. Biologists do not discover any law of universal ruthless selfishness but certain patterns of causes and effects within the organic world. A phrase such as 'universal ruthless selfishness' is, thus, at best a metaphor or at worst an expression of confusion.

So the first premise of the argument does not merely contain scientific information but also an extra-scientific value judgment and, therefore, Dawkins' argument against belief in God cannot be a scientific argument. This is not to say that the argument is a bad argument; in fact, it is one of the more forceful arguments, but it is nevertheless not a scientific argument. This indicates, however, that evolutionary biology can provide us with new information, which *in conjunction with* extra-scientific claims or value judgments can undermine existing religious beliefs. So scientific theories and discoveries can be of great relevance to religious belief. (But this is, of course, true only if T1 is false. But more on that later.)

A similar conclusion follows in respect to claim (4) and conclusion (5), which constitute Dawkins' positive argument for naturalism because (4) also contains a value judgment about evil states of affairs. But again, it remains true that evolutionary biology can provide us with new information, which in conjunction with extra-scientific claims or value judgments can support naturalistic beliefs. (I shall come back to this argument in the following section, when dealing with questions of the meaning or purpose of life and the universe.)

Of course, religious believers can dispute both of Dawkins' extra-scientific or philosophical arguments. Let me, very briefly, suggest two ways in which they could respond to the first argument. The reason why I am doing this is to show that premise (2) also requires the support of certain extra-scientific beliefs to have evidential force. Why should we accept the second premise of the argument, 'A good and powerful God would not allow evil and suffering of such a magnitude as we can find in the natural world'? The reason is presumably that if *we* cannot find any reason such a God might have for permitting evil of the magnitude found in the world, then probably such a God does not have any reason. Therefore, there probably is no God since we cannot find any such reason. Theists can respond in at least two ways to this line of argument. They can either claim that there are actually such reasons or question whether we are in a position to determine whether or not God has such reasons.

Why think that if God has reasons for permitting the amount of evil found in the world, we should be in a position to know this? After all we know from our experience that the epistemic distance between parents and say their 1-year-old child is so great that this child lacks the ability to understand most of the things the parents do. But, of course, religious believers think that the epistemic distance between a God who could create the heavens and the earth and human beings is at least as big as that between these parents and their 1-year-old child. Plantinga makes a similar point by asking us to compare two different situations:

> I look inside my tent: I don't see a St. Bernard; it is then probable that there is no St. Bernard in my tent. That is because if there were one there, I would very likely have seen it. ... Again, I look inside my tent: I don't see any noseeums (very small midges with a bite out of all proportions to their size); this time it is not particularly probable that there are no noseeums in my tent – at least it isn't any more probable than before I looked. The reason, of course, is that even if there were noseeums there, I wouldn't see 'em; they're too small to see.
>
> (Plantinga 2000: 466)[10]

Plantinga's claim is that the advocates of Dawkins' argument or of similar ones are in the latter kind of situation and they, therefore, are not justified in assuming that because they cannot see what reason a Christian God could have for permitting evil they are entitled to conclude, first, that such a God does not have a reason for permitting evil and, second, that probably such a God does not even exist.

Others have claimed that we can perhaps imagine some reasons God could have for permitting a great magnitude of evil in the world (see Hick (1966), Swinburne (1996) and Ward (1996)). Suppose a central idea behind the

---

[10] For a defence of this line of argument see also Wykstra (1990).

creation of the world is that God seeks to bring forth moral and spiritual beings who are capable of freely exercising faith in God, love towards their fellows and respect towards nature. So God is not trying to create a hedonistic paradise but an environment which makes possible the development of moral and spiritual virtues. But a world fit for human beings to develop the highest moral and spiritual virtues must, it seems, be one that includes the real possibility of suffering, disappointments, misery and disaster. It is because people can suffer harm, by such events as violence, accidents and starvation, that our actions affecting one another have real moral significance. Moreover, such an environment must operate according to general and dependable laws, for only on the basis of such laws could persons engaged in a spiritual and moral enterprise develop these qualities in a rational way. Much natural evil is, thus, a byproduct of the general laws of nature. So if God wants to make possible significant moral growth and development it seems as if God cannot create a world without moral and natural evil states of affair. God would then have a reason for permitting evil. Dawkins might, of course, still wonder if God could not have created a world with less natural evil than is actually the case. Perhaps, or then again perhaps not, because on this point it seems as if religious believers can appeal to some scientific discoveries to support their case. Cosmologists have pointed out, as we have already seen, that the fundamental physical constants and laws (such as the speed of light, the gravitational constant, the strong and weak nuclear forces) need to be exactly or almost exactly what they are for life and in particular for human life to be possible. But if this is the case, and God wanted to bring forth sentient beings with moral and spiritual virtues, then the fundamental physical constants and laws would have to be very much as they are, and they would as a consequence, it seems, necessarily involve all the possibilities of the suffering and evil we can find in the natural world.

My aim in this context, however, has been not to determine whose arguments (Dawkins' or his opponents') are most plausible, but to point out that this discussion – a very important discussion by the way – cannot take place within the sciences.[11] The reason is that the first premise of Dawkins' argument contains not merely scientific information but a moral evaluation of the natural processes that scientists can discover using biological methods. We are now, however, in a position to see that philosophy is needed to warrant the acceptance of the second premise as well because we need to consider whether the assumption it is based on (that if we cannot understand what reason God could have for permitting evil then probably God does not have a reason) ought to be accepted. Furthermore, we need to consider whether the reasons that God could possibly have for permitting this magnitude of evil, proposed by believers, are reasonable. Nothing, of course, would or should

---

[11] See Howard-Snyder (1996) for an interesting collection of articles on this theme.

stop Dawkins from participating in such a debate but he cannot fall back on the authority of science in doing so. Science alone does not undermine the compatibility of the existence of natural evil and a belief in a good and powerful God.

Evolutionary theory, chance and divine purpose

A number of biologists seem to think that evolutionary theory implies a meaningless universe. The idea is that biology somehow shows that life lacks a purpose and direction. The universe and its inhabitants are the result of chance and nothing more. Let me give some examples of biologists maintaining this view. (Some of these quotations have appeared before.) Gould tells us that 'Darwin argues that evolution has no purpose. Individuals struggle to increase the representation of their genes in future generations, and that is all' (Gould 1977: 12). William Provine asserts, 'Modern science directly implies that there... is no ultimate meaning for humans' (Provine 1988: 28). Dawkins maintains, 'The universe we observe has precisely the properties we should expect if there is, at bottom, no design, no purpose, no evil and no good, nothing but blind, pitiless indifference. ... DNA neither knows nor cares. DNA just is. And we dance to its music' (Dawkins 1995a: 133). Wilson writes, 'no species, ours included, possesses a purpose beyond the imperatives created by its genetic history' (Wilson 1978: 2). Lastly, George Gaylord Simpson claims, 'Man is the result of a purposeless and natural process that did not have him in mind' (Simpson 1967: 345).

But if Jews, Christians and Muslims are right that the universe is created by God and that God even intended to bring into being creatures made in God's image, then it seems as if the universe and human life have a purpose even if religious believers have not always been able to agree on what exactly it might be. But let us assume that Ward is right in that central to God's purpose is the 'generation of communities of free, self-aware, self-directing sentient beings' (Ward 1996: 191). Hence, on such a theistic account the purpose of genes is to build bodies, the purpose of bodies is to build brains and the purpose of brains is to generate consciousness and even self-consciousness, and with it appears for the first time in natural history, reflective and critical thinking, experiences of meaning, love and forgiveness and a capacity to choose between good and evil.

So there seems to be a serious clash between science and religion on this point. The theologian John F. Haught thinks that if these biologists are right then the conflict is so serious that 'although theology can accommodate many different scientific ideas, it cannot get along with the notion of an inherently purposeless cosmos' because such an idea is so central to a theological and religious concern (Haught 2000: 26). He does not think this is true merely of the major theistic religions but of most religions of the world. Haught writes,

> Since for many scientists today evolution clearly implies a meaningless universe, *all religions* must be concerned about it. Evolutionists raise questions not only about the Christian God but also about notions of ultimate reality or cosmic meaning as these are understood by many of the world's other religious traditions. ... Almost all religions, and not just Christianity, have envisaged the cosmos as the expression of a transcending 'order,' 'wisdom,' or 'rightness,' rather than as an irreversibly evolving process. Most religions have held that there is some unfathomable 'point' to the universe, and that the cosmos is enshrouded by a meaning over which we can have no intellectual control, and to which we must in the end surrender humbly.
>
> (Haught 2000: 9)

So there are good reasons why traditional believers of all sorts ought to take seriously these claims made by scientists and in particular by evolutionary biologists. The claims made are essentially: (1) evolutionary theory implies a meaningless universe, that is, that there is no ultimate meaning or that the universe is not here for a reason; (2) evolutionary theory implies, more specifically, that there is no meaning to be found behind the emergence of human beings in natural history, that is, we are not here for a reason and in particular we are not planned by God. In short, the key assertion is that *evolutionary theory undermines the religious belief that there is a purpose or meaning to the existence of the universe and to human life in particular.* It shows that the universe and humans are not here for a reason.

Notice that some of the quoted statements above seem to suggest yet another way evolutionary theory can undermine religious belief. A traditional religious understanding of the world involves the idea that a central part of the meaning of human life should be to truly love and respect God and other humans (and, as some like to add, other living creatures as well). But what Gould writes seems to conflict with such understanding of the meaning of life: 'Darwin argues that evolution has no purpose. Individuals struggle to increase the representation of their genes in future generations, and that is all' (Gould 1977: 12). This indicates that if there is any purpose to human life it consists in maximizing one's fitness. Dawkins is, as we have seen, very explicit on this point. He believes that science and in particular biology has a great deal to say about the meaning of life. It tells us that 'we are machines built by DNA whose purpose is to make more copies of the same DNA ... That is *exactly* what we are for. We are machines for propagating DNA, and the propagation of DNA is a self-sustaining process. It is every living object's sole reason for living' (quoted in Poole 1994: 58). This is what Dawkins, in his more recent writing, calls the 'single Utility Function of life', and he believes that 'everything makes sense once you assume that DNA survival is what is being maximized' (Dawkins 1995a: 106). The purpose of our human lives is, thus, to be survival machines for our genes and we serve this purpose best by maximizing our offspring. But if the meaning of life is to increase the representation of one's genes in future generations, and that is all, then this means that evolutionary theory undermines the religious belief that the meaning of life is to be found in a loving relationship with God and with other human beings.

So evolutionary theory is taken to undermine two separate religious claims about the meaning of life. The first deals with the meaning *of* life and the second with the meaning *in* life.[12] The former is a claim about whether the universe and life have any overarching purpose or ultimate meaning. The latter is a claim about what particular values and interest we ought to structure our lives around to give them meaning. One could claim, and this is what Dawkins, at least, seems to be doing, that there can be a meaning in life without there being a meaning *of* life. So even if our life cannot have any ultimate meaning it can still have a meaning, but it is an anti-religious meaning in that it is restricted to the activity of increasing the representation of our genes in future generations.

So what should we say about these claims? The second way in which evolutionary theory is taken to undermine religious belief seems obviously false, especially in the light of our discussion about the relevance of biology for morality and ethics. Moral activities and convictions exist that do not favour, and even hinder, the survival and reproduction of the individuals and their genes. Therefore, there is no reason to believe that we are in any significant way survival machines for our genes. This applies to the human search for meaning as well. We have a freedom to give our life meaning by structuring it around values other than maximizing fitness, such as appreciation of beauty and music, friendship and moral virtues, and interests such as football, sailing, science, caring for the poor and travel. By letting our activities be guided by values and interests such as these, we develop a reason or purpose sufficient to give our life a meaning. And, of course, among these values and interests a relationship with God can have a central place. So there are good reasons to believe that it is false that evolutionary theory undermines the religious belief that the meaning of life is to be found in a loving relationship with God and with other human beings.

But without God (or a sacred transcendent reality) the chances that there is an ultimate meaning or a meaning *of* life seem very slim indeed. Has evolutionary theory shown that there is no reason to hope for such an overarching meaning? In particular, can a science such as biology really demonstrate that it is not rational anymore for well-informed persons to believe that God brought the universe into being in order to realize a set of values or worthwhile states, including, in particular, the emergence of a complex self-conscious life form such as *Homo sapiens*?

It is not so easy to determine what exactly the argument is that these

---

[12] See Holmberg (1994) and Stenmark (1995b) for different attempts to develop this distinction.

biologists appeal to, to justify their claim that evolutionary theory undermines the religious belief that there is a purpose or meaning to the existence of the universe and to human life in particular. The conclusion is more often stated than the premises that warrant such a conclusion. But it seems to have something to do with the fact that evolutionary biologists have discovered that central to the development of life is chance or randomness. Dawkins writes, as we have seen, that 'natural selection, the blind, unconscious, automatic process which Darwin discovered, and which we now know is the explanation for the existence and apparently purposeful form of all life, has no purpose in mind' (Dawkins 1986: 5). But, of course, theists are not committed to believe that natural selection had any meaning in mind simply because natural selection is not an agent and as far as we know only agents can have purposes in mind. What they are committed to believe is that *God had a purpose in mind in using natural selection as a means to create human beings and that we, therefore, exist for a purpose*. The question is then whether science undermines such a religious belief. To be able to argue that that is the case, it seems as if one must show that *natural selection* and *God bringing us intentionally into existence* are incompatible beliefs.

What Gould writes may prove to be a good starting-point for such an argument because he maintains that evolutionary biology has shown that 'we are the accidental result of unplanned process ... the fragile result of an enormous concatenation of improbabilities, not the predictable product of any definite process' (Gould 1983: 101–2). That is to say, evolutionary biologists cannot find any propensities in the organic material they investigate that make the development of conscious life forms likely. On the genetic level all they find is random genetic changes (such as mutations) which are unrelated to the needs of the organisms. Natural selection operates over these chance events, but the selection is for survival and not increased biological complexity, which is necessary for human life to develop. It is, thus, a random event that evolution has led from simple life forms to complex life forms because such development is not probable given the biological mechanisms we know are at work in the evolutionary process, or at least the development of a self-conscious life form is not likely given these mechanisms. Hence, all biological events taking place in evolutionary history, including the emergence of *Homo sapiens*, are random with respect to what evolutionary theory can predict or retrospectively explain. Consequently, evolutionary theory has shown that human beings are *merely* accidental. There was not any plan, any foresight, any mind, or any mind's eye involved in their coming into being. Therefore, there is no ultimate meaning and in particular human beings are not here for a reason. There is no divine purpose to be found behind the emergence of life and of a self-conscious life form in particular.

What should we say about this argument? To start with it does not seem as though everything in the universe is really random because the things that exist obey general laws. It is not the case that everything is chaotic in the sense that anything could happen in natural history. If so one would expect that the natural laws would not be describable by science because they would change all the time or simply cease to exist. That this is not so is, of course, nothing that surprises religious people who believe that God planned the creation of the universe. Moreover, and as we have already seen, scientists have discovered that even a marginal difference in the initial conditions of the universe would have ensured that no life ever evolved anywhere. For life to emerge at all, the rate of expansion of the universe, the force of gravity, the weak and strong nuclear forces, and innumerable other physical conditions had to fall within a very narrow range. If not, the universe could never have produced hydrogen atoms, supernovae, carbon, water and other elements essential to life. The universe seems to be fine-tuned to support intelligent life. So even if the biologists are right that natural selection is a random process in the sense that evolutionary theory cannot predict or retrospectively explain its outcome, once we bring in physics and cosmology things appear to change. We then learn that the physical constants and initial conditions of this universe are suited with great precision to the evolution of living beings. It is no longer very surprising that evolution would eventually bring conscious life into existence.

I do not think that the accounts of evolutionary biologists, on the one hand, and the accounts of physicists and cosmologists, on the other, need to come into conflict with each other. The reason is that biologists typically focus on the evolution of a particular lineage of animals – it could have developed in a number of quite different ways from the way it actually developed – and the types of life forms and functions served. Rolston writes,

> Assuming more or less the same Earth-bound environments, if evolutionary history were to occur all over again, things would be different. Still, there would likely again be organisms reproducing, genotypes and phenotypes, natural selection over variants, multicelluar organisms with specialized cells, membranes, organs; there would likely be plants and animals: photosynthesis or some similar means of solar energy capture in primary producers such as plants, and secondary consumers with sight, and other sentience such as smell and hearing; mobility with fins, limbs, and wings, such as in animals. There would be predators and prey, parasites and hosts, autotrophs and heterotrophs, ecosystemic communities; there would be convergence and parallelism. Coactions and cooperations would emerge. Life would probably evolve in the sea, spread to the land and the air. Play the tape of history again; the first time we replayed it the differences would strike us. Leigh Van Valen continues: 'Play the tape a few more times, though. We see similar melodic elements appearing in each, and the overall structure may be quite similar. ... When we take a broader view, the role of contingency diminishes. Look at the tape as a whole. It resembles in some ways a

symphony, although its orchestration is internal and caused largely by the
interactions of many melodic strands.'

(Rolston 1999: 20)

So perhaps it is true that the development of *Homo sapiens* is not likely given
the scientific theories we have, but the development of some form of
intelligent life is. If we play the tape again and again it is likely that something
like us will appear.

So far we have discussed what appears to be the first premise in the 'no-
cosmic-purpose argument', namely the claim that we are the accidental result
of an unplanned process, a fragile result of an enormous concatenation of
improbabilities, and not the predictable product of any definite process. We
have seen that it is quite plausible to maintain the opposite conclusion that we
(or something similar to us anyway) are a predictable product of evolution
given the physical constants and initial conditions of the universe. That is to
say, if we state the first premise of the argument as,

(1) All individual species that come into existence through the process of
    evolution are random (that is, have a low probability) with respect to
    what evolutionary theory (or more broadly the sciences) can predict or
    retrospectively explain

then (1) seems to be a scientific premise but it is, as we have seen, possible to
question it on scientific grounds. The conclusion the advocates of this
argument want to validate is

(2) Therefore, the existence of human beings lacks an ultimate meaning, in
    particular, their existence is not the result of a divine purpose or
    intention.

Or alternatively,

(2´) Therefore, the existence of human beings is the result of pure chance and
     nothing more.

But (1) in itself is not sufficient to warrant (2) or (2´), other premises are
needed. To obtain (2´) one could presuppose the truth of (T1) or (T2). If so
one would perhaps add the following kind of premise to the argument:

(3) The only things that exist are the ones science can discover (or at least
    there is no reason to believe that anything over and above what science
    can discover exists).

The argument would then be that if only the entities and processes that science
deals with are real and it is the case that only science can discover random
events, than it follows that the existence of human beings is the result of pure
chance and nothing more. But we have already seen in Chapter 2 that premise

(3) is an extra-scientific claim and, furthermore, we have found good reason to believe that this claim is false. In any case what we have here is an argument against traditional religion which already presupposes the truth of scientific naturalism and therefore begs the question.

The only other plausible way these scientific expansionists can, as far as I can see, undermine religious belief about an ultimate meaning to the universe and to human existence is to add to (1) a premise about the conditions that must be satisfied for something to exist for a reason or to be something which is intended or planned by an agent. Remember that the religious belief is that we are planned by God to be here, that there is in this sense a meaning or purpose to our existence. We are not merely accidental because God intended to create us and did so, we have discovered, not by a direct act of creation but by the process of evolution. It seems, however, as if a requirement for a plan, purpose, foresight or intention to be involved in an object coming into being is that this object is not the result of pure chance, but has a certain likelihood of obtaining.

In our garden there are a number of flowers, some of which were planned by my wife and me to be there. There is, therefore, a reason for those flowers being there. Suppose that a friend said that because my wife and I planted a number of red roses and intended them to be there, there is a purpose according to which they grow in our garden. But there would not be a purpose if the reason why the red roses are there is simply that I tripped and some seeds accidentally lying on my sleeve merely happened to be thrown up in the air and happened to land in the flowerbed. So intention is necessary, but it is not sufficient for purposive outcome of action. Suppose I intended to plant red roses but I did this by simply taking topsoil out of a bag and putting it in the flowerbed. Accidentally there happen to be in the bag some red rose seeds, so after a while red roses start to grow where I intended red roses to grow. In this very hypothetical case we would not say that the roses were put there intentionally, rather that their presence is merely accidental. So intention plus a certain probability that the event intended actually obtains appears to be required for the purposive outcome of an action.

If the defenders of the no-cosmic-purpose argument apply these observations about human agents to God, it seems as if they have a complete argument. It would go something like this:

(4) The existence of *Homo sapiens* is planned by God only if the species' existence is intended by God and it is likely that its emergence will take place for that reason.
(5) But all individual species that come into existence through the process of evolution are random (that is, have a low probability) with respect to what evolutionary theory (or more broadly, the sciences) can predict or retrospectively explain.

(6)   Therefore, the existence of human beings lacks an ultimate meaning and, in particular, their existence is not the result of God's purposes, intentions or plans.

But there are, at least, two additional problems with this argument, besides the problem we have already discussed concerning premise (5). The first problem is that premise (4) is not a scientific premise but rather an extra-scientific or philosophical one, and it is also a premise that needs to be supported by philosophical arguments. This does not mean that the argument is necessarily a bad argument. I personally think it is an interesting one, but it is nevertheless – and this is what is important in this case – not a scientific argument. Hence, it is not true that science (or evolutionary biology) itself undermines the religious belief that there is a purpose or meaning to the existence of the universe and to human life in particular. Science cannot establish that the universe and humans are not here for a reason. What is true is that scientific theories such as evolutionary theory can *in conjunction* with extra-scientific or philosophical claims undermine such a religious belief. Note, however, that to the extent one thinks that this is possible, to that extent one also undermines the plausibility of T1. This is so because such extra-scientific premises are, of course, not species of scientific knowledge. But to have force they must be considered to be true, that is, someone (or many ideally) either knows them to be true or is at least rationally entitled to believe them to be true. So if we take these extra-scientific claims seriously we also by doing so question the truth of T1.

The second problem concerns premise (5). If one wants to obtain the conclusion that our existence lacks an ultimate or divine purpose and, in particular, that we are not intended to be here by a God such as the theists believe in, then what premise (5) tells us is not strictly relevant. The relevant issue is not, strictly speaking, what is likely given the scientific information or theories we possess, but what is probable given *God*'s knowledge about the outcome of the evolutionary processes that science investigates, if certain initial conditions are initiated at the beginning of the universe. Theists agree that such a being's cognitive capacity would outrun our capacity by far. They disagree, however, whether God's knowledge includes merely what *has* occurred and *is* occurring, or if it also includes all that *will* occur. Some theists even think that God possesses 'middle knowledge', that is, God also knows what *would* in fact happen in every possible situation or possible world (see, for instance, Hasker (1989), Zagzebski (1991) and Basinger (1996)). But even if God's knowledge is limited to everything that is or has been and what follows deterministically from it, it seems as though God's ability to predict with great accuracy the outcome of future natural causes and events is enormous. We cannot, therefore, automatically assume that what is likely given such divine knowledge is the same as what is likely given the scientific knowledge that we happen to have. So if God planned to create us and if it is likely that we would actually come into existence, given what God can know about the future of the evolving creation, then one could reasonably claim that we are here for a reason, and that in this sense there is a purpose to our existence. To establish the opposite conclusion seems to require more than basing one's calculation of probable outcomes on current scientific theories. At any rate, it follows that a successful defence of the relevance of premise (5) takes us far outside the domain of science and into metaphysics and theology. So any inferences from evolutionary biology that the universe or human existence is purposeless cannot possibly be categorized as scientific.

To sum up our discussion, we have considered the issue whether scientific theories and evolutionary theory in particular undermine religious beliefs, which might even be crucial for traditional religions such as Judaism, Christianity and Islam. This undermining might be taken as a reason to reject these religions and this is typically the way scientific expansionists and scientific naturalists in general understand the matter, but it can also be taken as an opportunity to reform and revitalize a religion. Haught takes this second road in *God after Darwin* and he even talks about evolutionary theory as 'Darwin's gift to theology' (Haught 2000: 45). Although the latter possibility is worth bearing in mind, our overall focus leads us in a different direction because it has been directed at attempts to expand the boundaries of science in different ways and at scientists who strive to undermine religion in order to replace it with naturalism. Our conclusion is that science indeed can be of relevance to religion in the sense of undermining (or supporting) traditional religious beliefs, but in the cases we have considered it does not do this in a direct way. That scientists like Dawkins, Gould, Provine and Wilson think that science alone is sufficient to refute beliefs central to traditional religions is the result of (a) misunderstanding religion (treating it as a scientific hypothesis); (b) misunderstanding science (conflating science and naturalism); or (c) not being sufficiently aware that the scientific theories they base their arguments on also require additional support of extra-scientific or philosophical premises to validate their conclusions.

But even though science cannot directly undermine traditional religions, can it still provide an alternative to them? Let us now turn to this issue.

Having science as one's religion

Can science compete with traditional religions and what is the content of this new scientific faith? Wilson, as we have seen, believes that science has shown or will, at least in a near future, show that traditional religious beliefs are really enabling mechanisms for survival and apparently nothing more. In the place of religion Wilson thinks we should put something he variously styles 'scientific materialism', 'scientific naturalism' or 'scientific humanism'. (In what follows I will merely use 'scientific naturalism' to denote this position.) Wilson believes that scientific naturalism 'presents the human mind with an alternative mythology that until now has always, point for point in zones of conflict, defeated traditional religion' (Wilson 1978: 192). He also adds that the best scientific theory to base one's scientific mythology or religion on is evolutionary theory: 'the evolutionary epic is probably the best myth we will ever have' (201).

Wilson does not differentiate between science, on the one hand, and scientific naturalism, on the other, hence he seems to think that there is no reason to uphold any distinction between them. Consequently, science can be understood as – or legitimately expanded into – a mythology. This I shall interpret to mean, at least, two things. First, that what is believed cannot be proven. One can perhaps give good reasons for their truth but never conclusive ones. Wilson explicitly affirms this because he writes,

> The evolutionary epic is mythology in the sense that the laws it adduces here and now are believed but can never be definitely proved to form a cause-and-effect continuum from physics to the social sciences, from this world to all other worlds in the visible universe, and backward through time to the beginning of the universe.
>
> (Wilson 1978: 192)

Second, it must be a response to our existential concern. A mythology *per se* must deal with ultimate questions concerning who we are, why we exist and what the meaning of our life is, and what stance we should take towards experiences of death, suffering, guilt, anxiety, love and friendship, and the like.

Wilson is then what we might call a 'science believer', in contrast to 'religious believers' and 'ideology believers'. A *science believer* is someone who believes that science alone can solve our existential or ultimate questions or at least can do this better than any traditional religion or secular ideology. A science believer thinks that science can be one's religion in the functional sense. Religious believers and ideology believers, however, deny this. Science can provide us with information that is relevant when dealing with religious or ideological matters, but science cannot itself be a religion or ideology. Religious believers and ideology believers differ in the sense that the former are convinced that the sacred or God provides the way to solve our existential problems, whereas that is something the latter deny (or at least are agnostic about).

Is Richard Dawkins also a science believer? The answer seems to be 'yes'. Let me repeat some earlier quotations from Dawkins to verify this. Dawkins claims that science, contrary to what many people have thought, has a great deal to say about our existential questions:

So where does life come from? What is it? Why are we here? What are we for? What is the meaning of life? There's a conventional wisdom which says that science has nothing to say about such questions. Well all I can say is that if science has nothing to say, it's certain that no other discipline can say anything at all. But in fact science has a great deal to say about such questions.

(quoted in Poole 1994: 57)

He, furthermore, tells us that since we have modern biology, we have 'no longer ... to resort to superstition when faced with the deep problems: Is there a meaning to life? What are we for? What is man?' (Dawkins 1989: 1). Science is capable of dealing with all these questions and constitutes in addition the only alternative to superstition. He quotes and agrees with G. G. Simpson, who writes that 'all attempts to answer that question [What is man?] before 1859 are worthless and that we will be better off if we ignore them completely' (Dawkins 1989: 1). What answers does science then give to these questions? Science tells us that 'We are machines built by DNA whose purpose is to make more copies of the same DNA ... That is *exactly* what we are for. We are machines for propagating DNA, and the propagation of DNA is a self-sustaining process. It is every living object's sole reason for living' (quoted in Poole 1994: 58). Dawkins maintains that the evolutionary theory can explain not merely who we are, but also why we exist and what the purpose of our life is.

Dawkins, like Wilson, thinks that science and religion compete, not merely because science can be expanded to deal with our existential concern but because he believes, as we have seen, that traditional religions are actually scientific or quasi-scientific hypotheses or theories (Dawkins 1995b: 46–7). Dawkins, thus, raises a scientific challenge to traditional religions, which he thinks they cannot meet. In fact, evolutionary theory makes it possible to be an 'intellectually fulfilled atheist' – something which Dawkins think was not really possible before Darwin (Dawkins 1986: 6).

Note, however, some differences on this point between Dawkins and Wilson, on the one hand, and Ruse, on the other. Michael Ruse, Wilson's co-writer on many occasions, does not accept the idea that science can replace traditional religions, and thus constitute our new religion. In the new, last chapter of the second edition of his book *Taking Darwin Seriously*, Ruse writes,

If you buy the chief message of this book, you are going to accept a naturalistic account of both epistemology and of ethics. If, like me, you are a sceptic, not knowing if anything lies beyond, then that is all you are going to get. You do not have a religion, but you have something instead. If you have a religion as well, then so be it. You can fuse your Darwinism onto it. A Johnsonian contradiction between science and religion is not inevitable. But neither is a Dawkinsian contradiction between science and

> religion. Most especially, a 'Darwinist religion,' in the sense that
> Darwinism is the religion, does not have to be part of one's package.
>
> (Ruse 1998: 294)

Ruse disagrees then with both Dawkins and Wilson about whether science can
constitute a religion. Whether one should have a religion or not is a question
of importance but one that Ruse thinks falls outside the scope of the sciences.

   Ruse seems in these remarks to allow for a different way of understanding
the relationship between science and religion than that proposed by Dawkins
and Wilson. Dawkins and Wilson assume what we can call a *unity model*,
saying that science and religion are competitors who try to deal with the same
kinds of questions and explain the same range of phenomena. Other available
options seem to be either an intersection model or a separation model. The
intersection model says that in some areas there is an overlap between science
and religion. Intersectionists may disagree about exactly where there is such
an overlap, but they, nevertheless, agree that since there is only an intersection
between science and religion and not a union, science cannot replace religion
or vice versa. A reason typically given why there is not a union is that religion,
but not science, deals with our moral and existential concerns. Thus, science
can be of relevance in the development or reconstruction of a religion (or an
ideology) but cannot itself become a religion (or an ideology). The third
logical alternative, the *separation model*, is defended by those who maintain
that science is not in any significant way relevant for the development or
reconstruction of a religion (or an ideology) because these practices do not
overlap in any area of importance.

Scientific naturalism and philosophical naturalism

The difference between Dawkins and Wilson, on the one hand, and Ruse, on
the other, indicates that we have to distinguish between two versions of
naturalism. What are the differences between these two versions? Let us begin
by noticing that naturalism, according to Kai Nielsen, 'denies that there are any
spiritual or supernatural realities. There are, that is, no purely mental
substances and there are no supernatural realities transcendent to the world'
(Nielsen 1997: 441). Naturalism is, thus, opposed to traditional Judaism,
Christianity and Islam as well as Buddhism (which has neither God nor
worship, but a belief in spiritual realities). Naturalism is therefore not agnostic,
but a form of atheism (442). Positively stated naturalists claim that matter or
physical nature is ultimately real in the sense that everything that exists is
either merely physical or cannot exist without a physical basis. Let us call this
aspect of naturalism the *metaphysical element* because it makes claims about
what ultimately exists and the relationship between these entities.

   Nielsen maintains, moreover – and this is important for our concern – that naturalism need not be 'reductionistic (claiming that all talk of the mental can be translated into purely physicalist terms) or scientistic (claiming that what science cannot tell us humankind cannot know)' (Nielsen 1997: 440). Instead Nielsen believes that the 'more plausible forms of naturalism are neither across the board reductionistic nor scientistic', but non-scientific (440). Hence, Nielsen's naturalism entails a rejection of T1, namely, the thesis that the only kind of knowledge we can have is scientific knowledge. This makes his naturalism different from Dawkins' and Wilson's because they in their argument seem to presuppose that our knowledge is limited to what can be obtained by using the methods of science. In other words, scientific expansionists like Dawkins and Wilson add to the metaphysical element also an *epistemological element*. Notice, however, that the metaphysical element includes also an additional claim to the one about what ultimately exists, namely one about the relationship between the physical and the mental. Nielsen's naturalism asserts that mental states are real but causally dependent on physical states. The personal and mental, though distinct from the physical, are fully caused by the latter. The mental supervenes on the physical. Nielsen, therefore, takes his naturalism not to be reductionistic. But we have seen that at least some scientific expansionists maintain a form of naturalism that is reductionistic in this sense. Crick's 'astonishing hypothesis', for example, is that we are nothing but a pack of neutrons (Crick 1994: 3).

These differences imply that it is important to distinguish the scientistic form of naturalism from the one advocated by Nielsen and others. Let us call the former *scientific naturalism* because its defenders claim that naturalism can be scientifically justified and hence be a part of science. Those like Nielsen who accept a *philosophical naturalism* maintain instead that their naturalism is inspired by science or based on scientific results among others, but justified by philosophical arguments and therefore not a proper part of science. Philosophical naturalism is informed by science, but nevertheless transcends it in content. Thus, philosophical naturalists accept some kind of scientific restrictionism and an intersection model of the relationship between science and religion, whereas scientific naturalism is an expression of scientific expansionism and assumes a unity model.

There is one more difference between scientific and philosophical naturalism worth mentioning in this context. Both scientific and philosophical naturalists agree that 'people can make sense of their lives and reasonably order their lives as moral beings without any belief in God or any other spiritual realities' (Nielsen 1999: 443). Let us style these the *axiological* and *existential elements*. The naturalists maintain that it is possible to create meaning in life, justify moral convictions and live a good life without belief in God. But whereas many scientific naturalists believe that science can deliver what is needed in this case, philosophical naturalists think that moral and existential concerns fall outside the scope of science.

The advocates of scientific naturalism assert, more exactly:

(1) Only science can provide us with knowledge about reality.
(2) Matter in motion or physical nature is what is ultimately real.
(3) Mental states can be completely reduced to physical states (or at least mental states are causally dependent on physical states).

These assertions are I suggest the core claims of scientific naturalism. To them are sometimes added, as we have seen, the assertions:

(4) Science can tell us how to behave morally or how to deal with our moral concern.
(5) Science can tell us what is the meaning of life or how to deal with our existential concern.

But claims (4) and (5) are in this particular case crucial because without them scientific naturalism could not be a competitor to traditional religions and secular ideologies. Science cannot be one's religion or a mythology without those last two assertions.

What our discussion of Scientism suggests is that it is typically not the metaphysical claims but the epistemological claim that constitutes the key element of scientific naturalism, in contrast perhaps to philosophical naturalism. It is by applying scientific methods (that is, the methods of physics, chemistry and biology) that scientific naturalists attempt to justify claims (2), (3), (4) and (5). If claim (1), what we previously called T1, is the core of scientific naturalism then it is strictly speaking not incompatible with belief in God. All that claim (1) tells us is that God cannot be known unless God acts in observable ways in human affairs and, thus, can be discerned by the methods of science. But by using the methods of physics, chemistry, biology and the like, scientists have only discovered matter in different constellations in motion governed by natural laws and chance. There is, therefore, no reason to believe that what ultimately exists is anything else than matter in motion. Neither can we, by investigating things with these methods, discover that anything mental exists which is not completely reducible to the physical, or at least the sciences cannot support the idea that mental states can exist causally unrelated to physical states. Consequently, whereas claim (1) does not logically speaking exclude a divine reality, claim (3) precludes such a reality if it is understood as a spiritual reality which is causally independent of physical reality. Thus, a God who is understood to have a conscious life, and thus mental states without physical states cannot exist because mental states are always causally dependent on physical states. Hence, scientific naturalism precludes the classic theism of Judaism, Christianity and Islam (but, of course, not all conceptions of the divine in these traditions, since, for instance, process theism is still a possibility) (see for examples Barbour (1990) and Haught (2000)).

Scientists can, furthermore, investigate moral codes by using biological methods, which are discovered to be genetic mechanisms for maximizing fitness, and on the basis of this information tell us how we morally ought to behave, and also discover the only purpose of life that can be scientifically discerned, namely that our purpose is to be vehicles for propagating DNA and that that is every living object's sole reason for living.

It is important to take scientific naturalism (or scientific materialism) into account because it is by no means uncommon to adopt this viewpoint within the scientific community. Haught maintains that it 'has become so intimately intertwined with modern science that today many scientists hardly even notice the entanglement' (Haught 1995: 33). According to Paul M. Moser and J. D. Trout, 'Materialism is now the dominant systematic ontology among philosophers and scientists ... As a result, typical theoretical work in philosophy and the sciences is constrained, implicitly or explicitly, by various conceptions of what materialism entails' (Moser and Trout 1995: ix). Whether Haught, Moser and Trout are right about this is hard to determine, but at least it is true that we can find, as we have seen, influential scientists expressing these ideas. Sometimes this is done in subtle way by bonding their scientific theory construction with hidden premises of scientific naturalism. On other occasions, it is stated more openly.

Scientistic faith

The third way in which one could assume that science (or, in particular, evolutionary theory) can be of great significance for religion is then that it can replace traditional religions and provide us with a new religion, that is, scientific naturalism. What should we say about the prospects of such a view? *Ought we to become science believers, that is, to accept scientific naturalism as our new religion?* We face at this point essentially two issues. The first is whether scientific naturalism satisfies the requirements for being a religion or view of life, whether it can fulfil that particular function in human life. The second is whether scientific naturalism really is science, whether science legitimately can be expanded into a religion or view of life.

I have argued elsewhere that for something to be a view of life or a religion in the function sense it must satisfy certain requirements (Stenmark 1995a: 235–68). Is Scientism in the form of scientific naturalism able to do this? A view of life must fulfil, at least two tasks. First, it must structure and make reality intelligible (the *theoretical function of a view of life*). That is, it must to some degree make the world a cosmos and determine the place of human beings in it, and also state what is of value in life. Second, a view of life must concretely guide people in how they should live their lives, how they should deal practically with their existential experiences of, for instance, meaninglessness, suffering, guilt and love, and their interpersonal relationship with other human beings (the *regulative function of a view of life*). This is so because believing in a view of life is not just a matter of seeing the world in a particular way, but also a matter of choosing a way of living.

Scientific naturalism is able to fulfil the theoretical task. It can provide its adherents with a map of reality. It can tell us where human beings fit in and what the central values of our existence are. It is less certain whether scientific naturalism can concretely regulate people's lives in the way traditional religions have been able to do. Wilson himself seems aware of this problem. He writes that the 'fatal deterioration of the myths of traditional religion' has led to 'a loss of moral consensus, a greater sense of helplessness about the human condition and a shrinking of concern back toward the self and the immediate future' (Wilson 1978: 195). Scientific naturalism must face this challenge. It must supply people with a new myth powerful enough to overcome these destructive consequences of the deterioration of traditional religious myths. It must be able to provide a faith by which people actually could live, not only a theoretical map of reality. Scientific naturalism must not merely be a vision *of* life but be a vision *for* life. Wilson thus suggests:

> a modification of [traditional] scientific humanism through the recognition that the mental processes of religious belief – consecration of personal and group identity, attention to charismatic leaders, mythopoeism, and others – represent programmed predispositions whose self-sufficient components were incorporated into the neural apparatus of the brain by thousands of generations of genetic evolution. As such they are powerful, ineradicable, and at the center of human social existence. ... I suggest further that scientific materialism must accommodate them on two levels: as a scientific puzzle of great complexity and interest, and as a source of energies that can be shifted in new directions when scientific materialism itself is accepted as the more powerful mythology.
>
> (Wilson 1978: 206–7)

However, it is not possible now to predict the form religious life and rituals will take as 'scientific materialism appropriates the mythopoeic energies to its own ends' (Wilson 1978: 206). But Wilson admits that here lies at least the present 'spiritual weakness' of scientific naturalism. It lacks the 'primal source of power' that religion for genetic reasons is hooked up with, partly because the 'evolutionary epic denies immortality to the individual and divine privilege to the society' (192–3). Moreover, scientific naturalism will 'never enjoy the hot pleasures of spiritual conversion and self-surrender; scientists cannot in all honesty serve as priests' (193). But Wilson, nevertheless, believes that a way exists to divert the power of religion into the service of scientific naturalism, even if the future will have to tell us how exactly this will be done.

So it seems as if scientific naturalism can be or can at least have the potential to become a full-fledged view of life or religion, even if it would lack some of the attributes of traditional religions. But this would be something it would have in common with other views of life such as environmentalism, feminism and Marxism.

The real problem with this new religion or view of life is not whether it can fulfil the conditions for being a religion, but whether it is science or a proper part of science. I personally think the universe is religiously speaking sufficiently ambiguous, especially after the development of evolutionary theory, that naturalism is a defensible view of life (although I ultimately believe, but that is a different matter, that its central claims are false). But scientific naturalism, in contrast to philosophical naturalism, *pretends to be science* or *to be a necessary presupposition of science* or *to be directly implied by science* and this is something quite different. It is something we by now have good reasons to reject. We have seen that to work properly science needs to presuppose naturalism only in the methodological sense. Theories in chemistry, biology and physics can be true, and predict and explain events in the world, whether or not it is the case that matter or physical nature alone is ultimately real. Nor is it the case that science directly implies metaphysical naturalism in general or that science can show, for instance, that humans are nothing but matter in motion. Science cannot exclude the possibility that there are dimensions of reality that are neither describable in scientific language nor accessible to scientific explanations. To be able to do this science must be able to demonstrate that it gives an exhaustive account of reality. But this cannot be done in a non-question-begging way. The problem is that since we can only obtain knowledge about reality by means of scientific methods, we must use those methods whose scope is in question to determine the scope of these very same methods. If we used non-scientific methods we could never come to know the answer to our question, because according to scientistic faith there is no knowledge outside science. Furthermore, if, for instance, God exists, it seems almost uninformed to expect that such a reality should be graspable by means of scientific investigation and experimentation. If God exists, we would expect that God would not be that kind of being.

Crucial for a religion is also its capacity to tell its adherents what they ought to value, how they ought to behave towards other people and what interests they should develop and focus their attention on in life in order to live a morally good and meaningful life. But we have seen that this is not something science in general or evolutionary biology in particular can do. The attempt we have examined to derive what ought to be the case from scientific statements about what is or has been the case fails. Wilson does not succeed in his effort to show how the mammalian imperative, which he ultimately seems to think justifies universal human rights, follows from the mammalian plan evolutionary biologists believe they have discovered in nature. So scientific naturalism can become a religion, a view of life or mythology only if its advocates add certain extra-scientific claims to the scientific theories they base their beliefs on. It is the conflation of these elements that gives the false impression that science can be one's religion and in particular that evolutionary theory can be the best myth we will ever have.

All taken together means that scientific naturalists face a dilemma: *either* scientific naturalists maintain that what they are doing is science but then have to give up their missionary activities or their naturalism and become merely scientists, *or* scientific naturalists keep their naturalism but then have to admit that they are not doing science anymore. We ought, therefore, not to become science believers. Our existence – which Dawkins tells us is the greatest of all mysteries – is not a mystery that science can solve, although science can give us important information about it that we should not ignore. But we must find the answer to this mystery somewhere else.

In this chapter we have completed our discussion of Scientism and religion. We saw in the last chapter that the attempt by scientific expansionists to explain traditional religion as a wholly material phenomenon, whose ultimate function is to maximize genetic fitness and that religion, therefore, offers only mundane rewards, is not convincing. We have now reached a point where we can conclude that these scientific expansionists also fail to replace traditional religion with science. Science can, on the other hand, provide us with information that can undermine or support existing religious beliefs or view of life beliefs, but science cannot itself be a religion or a view of life properly understood. But we have also seen that when science offers such information, this information is seldom sufficient to warrant any direct conclusions concerning the truth or rationality of religious beliefs. To be able to do that philosophical and extra-scientific claims are also needed. So science is relevant for the development, the reconstruction and the replacement of religions or views of life, but typically not in the way Crick, Dawkins, Gould, Provine, Wilson and other scientific expansionists claim. Provine's observation that very few truly religious biologists remain might perhaps be correct (see p. 91), but he is certainly wrong in thinking that evolutionary theory itself justifies such a movement (Provine 1988: 28). Such a change in attitude towards religion follows only if biologists mistakenly conflate science and naturalism.

CHAPTER SEVEN

Scientism and the fear of religion: some concluding remarks

Our topic has been Scientism (or scientific expansionism) and the aim has been to clarify this view and assess critically the ideas it contains. We can say that what is characteristic for advocates of Scientism is that they believe that the boundaries of science (that is, the *natural* sciences) could and should be expanded in such a way that something that has not previously been understood as science can now become a part of science. How exactly the boundaries of science should be expanded and what more precisely it is that is to be included within science are issues on which there is disagreement. Some promoters of Scientism are more ambitious in their extension of the boundaries of science than others. In its most ambitious form Scientism states that science has no boundaries: eventually science will answer all our problems.

A common way of defining Scientism is to say that it is the idea that science tells us everything that there is to know about reality. It is an attempt to expand the boundaries of science in such a way that all genuine (in contrast to apparent) knowledge must either be scientific or at least be reducible to scientific knowledge. This view has in this study been named epistemic Scientism or merely T1. Sometimes epistemic Scientism is hooked up with ontological Scientism or T2. It is the view that the only reality that exists is the one science has access to. Only the things science can discover exist. Whereas epistemic Scientism does not entail ontological Scientism, ontological Scientism entails epistemic Scientism. Taken together we have the view that science sets both the limits of our knowledge and the limits of reality. We can say that scientists adhere to such a view when they present their scientific account as complete. Examples would be Crick's claim that we are nothing but a pack of neurons, Sagan's that the cosmos is all that is or ever will be and Simpson's that all attempts to answer the question 'What is man?' before Darwin's theory of evolution are worthless and that we will be better of if we ignore them completely.

Two other versions of Scientism that I have been especially interested in are axiological Scientism and existential Scientism. Axiological Scientism could be either the view that science is the only truly valuable realm of human learning or culture; all other realms are of negligible value, or the view that science alone can completely explain morality and replace traditional ethics. It is the latter view (labelled T3) that has been my main concern. Wilson's attempt to develop a biology of ethics would be a good example of such a view. Existential Scientism (or T4) is often related but nevertheless distinct from axiological Scientism (T3). It is the idea that science alone can explain as well as replace traditional religion. Science can answer our existential questions about who we are, why we exist and what the meaning of life is. Dawkins tells us, for instance, that the scientific answer to these questions is essentially that we are machines built by DNA whose purpose is to make more copies of the same DNA, and this is every living object's sole reason for living.

Axiological Scientism and existential Scientism must be distinguished from one another because it is possible to affirm, for instance, that evolutionary theory is the sole, or at least the most important, source for developing a moral theory and explaining moral behaviour, but at the same time to deny that biology or any other science can tell us what the meaning of human life is or that it can fulfil the role of religion in our lives. Thus, axiological Scientism does not entail existential Scientism. But does existential Scientism entail axiological Scientism? This is less clear, as we have seen. Religions and world views are in general taken to include some ideas about how we should live and about what constitutes a good human life. If this is correct then the acceptance of existential Scientism implies also an acceptance of axiological Scientism. But, on the other hand, it is perhaps possible to say that science alone can answer some of our existential questions and thus that science can partially replace religion.

The relation of these latter forms of Scientism to the earlier two is that neither axiological nor existential Scientism entail epistemic or ontological Scientism. It is not inconsistent to claim that science can answer our moral questions and replace traditional ethics or that science can answer our existential questions and replace traditional religion, without maintaining that the only reality we can know anything about or that the only reality that exists is the one science has access to. Nevertheless, it is true that the scientific naturalism that scientists like Dawkins and Wilson adopt as their world view or religion is often based on the previous acceptance of epistemic or ontological Scientism.

T1, T2, T3 and T4 state four of the ways (there are others as we have seen) in which Scientism can be either explicitly stated or implicitly assumed. I have suggested that it is sufficient that a person accepts one of these four theses to be counted as an adherent of Scientism. These four claims have also been assessed critically.

Since ontological Scientism or T2 entails epistemic Scientism or T1, the former cannot be true unless the latter is. Therefore, we have in the critical examination focused on T1. We have seen that there are domains of knowledge outside and independent of science. We have, at least, the domains of observation, introspection, self-reflection, memory, language and intention.

Scientific knowledge, furthermore, presupposes the existence of other reliable sorts of knowledge such as that derived from self-reflection, memory and language. So if we did not have these other kinds of knowledge, we would not be able to obtain any scientific knowledge at all. Let us here merely exemplify this with knowledge of memory.

To be able to develop and test a scientific hypothesis against a certain range of data, scientists have to be able to remember, for instance, the content of the hypothesis, the previous test results and more fundamentally that they are scientists and where their laboratories are located. Their scientific knowledge presupposes knowledge of memory. The truth is that unless we can trust our memories (and obtain knowledge), we can never reason at all nor do any science whatsoever, because in any inference we must remember our premises on our way to the conclusion. All activities we are engaged in therefore presuppose knowledge based upon memory. Accordingly, science presupposes the possibility and reliability of knowledge based upon memory. But if scientific knowledge is the only sort of knowledge we can have, then we cannot know anything at all. We would never be able to obtain scientific knowledge (or any other kind of indirect knowledge) unless we first had some knowledge based upon memory, nor for that matter could we remember that some people think that scientific knowledge is the only sort of knowledge we can have.

The most troublesome difficulty with T1, however, is that it is self-refuting. The problem is that the scientistic belief that we can only know what science can tell us seems to be something that science cannot tell us. How can one set up a scientific experiment to demonstrate the truth of T1? What methods in, for instance, biology or physics are suitable for such a task? Well, hardly those methods that make it possible for scientists to discover and explain electrons, protons, genes, survival mechanisms and natural selection. Furthermore, it is not because the content of this belief is too small, too distant or too far in the past for science to determine its truth-value (or probability). Rather it is that beliefs of this sort are not subject to scientific inquiry. We cannot come to know T1 by appeal to science alone. T1 is rather a view in the theory of knowledge and is, therefore, a piece of philosophy and not a piece of science. But if this is the case, then T1 is self-refuting. If T1 is true, then it is false. T1 falsifies itself.

Thus, T2 cannot be true because T1 is false. Let me nevertheless also mention a specific problem with T2: the view that the only reality that exists is the one science has access to. The difficulty is that its advocates often overlook a crucial distinction within scientific methodology, that between a methodological reduction and an ontological reduction. Scientists are justified in reducing a phenomenon (such as human beings) to another (such as a collection of neutrons) for a particular purpose, but without claiming that that is all this phenomenon is. Science as such allows only methodological and not ontological reduction. This is a wise methodological self-restriction because how could one demonstrate that science gives an exhaustive account of reality in a non-question-begging way? What we want to know is whether science sets limits for reality. The problem is that since we can only obtain knowledge about reality by means of scientific methods (that is T1), we must use those methods whose scope is in question to determine the scope of these very same methods. If we used extra-scientific methods we could never come to know the answer to our question, because there is according to Scientism no knowledge outside science.

Hence, there are good reasons to reject epistemic Scientism and ontological Scientism because both T1 and T2 have been shown to be philosophical rather than scientific claims. Therefore, our conclusion is that when scientists such as Crick, Dawkins, Sagan, Simpson and Wilson make their scientistic declarations, they are not speaking as scientists and are not doing proper science.

We have also examined critically axiological Scientism (or T3) and existential Scientism (or T4). We have found some influential advocates of these forms of Scientism within evolutionary biology. These scientists are ready to apply evolutionary theory to both ethics and religion. One problem though is that evolutionary theory can be claimed to be of relevance for ethics and religion in different ways. Therefore, we distinguished between the attempt to explain ethics or religion (project A), the attempt to provide us with new scientific information about life that undermines ethical views or traditional religious beliefs (project B) and the attempt to develop a scientific ethic or scientific religion, namely scientific naturalism (project C).

There is no doubt that project (A) falls within the scope of science. Once we accept that humans are also animals in the sense that they have a common ancestry with all other living things on this planet, then it must be appropriate to investigate to what extent explanations of the behaviour of animals can be applied to humans as well. But we have seen that the Darwinian explanation, that morality exists and continues to exist because it emerged and continues to function as a strategy (or part of a strategy) adapted to secure the fitness of the individuals and their genes, has limited explanatory scope. Vast stretches of the moral landscape cannot plausibly be explained as fitness-maximizing strategies. In fact, moral convictions exist, for instance in environmental ethics, that do not favour but even hinder the survival and reproduction of the individuals and their genes. A better explanation of much moral belief and behaviour is therefore that gradually, as we evolved from our pre-human ancestors, our brain grew and we began to reason to a degree no other animals had achieved, and it is the possession of this ability that make possible not only science, but also morality and our questions about which courses of action are for the best. These ideas about which courses of action are for the best are then spread non-genetically through cultural transmission or communication, and people have become convinced that slavery is morally wrong, that women ought to be given equal opportunities and responsibilities with men, that other living things besides humans have moral standing and so forth.

Project (B) is perhaps the most obvious way in which evolutionary theory can be of significance for ethics. The problem is, as we have seen, that philosophers are in general aware that well-informed moral judgments must not merely be based on appropriate ethical norms, but also on relevant factual information. But at least two of the claims made by scientific expansionists would, if true, have a profound impact on the conception of morality we find among philosophers and theologians. These are that morality is ultimately a matter of selfishness and that the objectivity of morality is an illusion. I have shown that the first claim is based on a conceptual confusion. These biologists do not clearly distinguish between selfishness and altruism in its moral and in its biological sense. The second claim fails because it undermines itself. The argument is that ethical norms or beliefs cannot be objective because they are merely the product of evolution. They are rather an adaptation put in place to further our reproductive ends and nothing else. But from the biological perspective science also is nothing else than a product of evolution. Thus, science cannot be objective, but is an adaptation put in place to further our reproductive ends and nothing else. But then there is no reason for us to believe that the objectivity of morality is an illusion because these biologists' claim is merely the product of evolution. In fact, if their theory is true (that our behaviour is firmly under the control of the genes and that we function better if we believe in the objectivity of morality), then it would be very unlikely that these biologists would be able to discover that the objectivity of morality is an illusion. So if these scientific expansionists are right, they are probably wrong.

Project (C) provides the most direct way in which evolutionary theory could be maintained to be of relevance for ethics. The claim is that evolutionary theory can itself justify ethical norms and beliefs and provide us with a new ethic. However, to be able to do this, scientific expansionists must · deal with the naturalistic fallacy. Wilson thinks that the naturalistic fallacy is something biology can refute because from the perspective of the natural sciences the ethical precepts are no more than principles of the social contract hardened into rules and dictates. Thus, ought-statements are just shorthand for one kind of factual statement and since science deals with factual statements it can also deal properly with moral statements. But we have seen that when we 'translate' ought-statements into is-statements in the way Wilson suggests, we can still not obtain any ought-conclusion telling us how we ought to behave or what we should do. Consequently, Wilson must get his ought-premises from some domain other than science, and thus the project of developing moral norms and justifying ethical beliefs does not genuinely fall within the scope of science, or more specifically within the scope of evolutionary biology.

Let us finally summarize our discussion about Scientism and religion. Can science explain as well as replace religion with scientific naturalism? Again, evolutionary theory can be of relevance to religion in the same three ways it can be of relevance to ethics.

Project (A), the attempt to develop a Darwinian explanation of religion as a purely material phenomenon whose function is to increase genetic fitness, has a certain plausibility dealing with tribal religion. But once a religion becomes universal, as many of the major world religions have done, this explanation will not do. Once outside the tribal context it is not plausible to maintain that religious believers, who are urging us to become Christians, Muslims or Buddhists, are doing so because they thereby increase their offspring in the next generation. Missionary behaviour of the magnitude present in many of the world religions provides powerful counter-evidence against a purely Darwinian explanation of religion. Religion cannot be explained as merely a fitness-maximizing strategy. A much more plausible explanation of these phenomena is that some religions have spread worldwide because they deal better than others with people's existential concerns and this information has spread non-genetically through cultural transmission or communication. Perhaps this is also the best scientific explanation of why religion exists and persists. Religion exists because it offers answers to questions about who we really are, why we exist and what the meaning of our life is, and what stance we should take towards experiences of death, suffering, guilt, anxiety, love and friendship. If this is correct, religion cannot successfully be explained as merely an illusion fobbed off on us by our genes to get us to cooperate in securing our personal and inclusive fitness. Successful religion is not really 'selfish' in the sense of serving one's genes. Religion is rather 'altruistic' in the sense that it typically offers its gospel to both relatives and non-relatives, slaves and free persons, male and female. So there is a clear limit to the extent evolutionary theory can explain religious belief and behaviour.

Science can clearly provide us with information that can undermine or support existing religious or life view beliefs; to this extent project (B) is successful. Hence, science has undermined the religious beliefs that the wind, rain and lightning are the direct manifestations of divine activity, that the earth was created in six days or six thousand years and that God created an original human pair, Adam and Eve. But we have also seen that when science offers such information this information is seldom sufficient to warrant any direct conclusions concerning the truth or rationality of religious beliefs. To be able to do that, philosophical and extra-scientific claims are also needed. So science is relevant for the development, reconstruction and replacement of religions or views of life, but typically not in the way Crick, Dawkins, Gould,

Provine, Wilson and other scientific expansionists claim. That these scientists believe these things is the result of either misunderstanding religion (treating it as a scientific hypothesis) or misunderstanding science (conflating science and naturalism).

Dawkins maintains, as we have seen, that belief in God is a competing explanation for facts about the universe and life, and therefore is a scientific hypothesis. I have argued that this is misleading. It is misleading because belief in God is not a theory or hypothesis invented to explain particular facts about the physical and biological world, as a scientific theory is, but is rather taken by religious believers to be the outcome of an encounter with a divine reality. A reality that these believers claim helps them in their lives to deal with experiences of suffering and anxiety, and which gives their lives a meaning. Dawkins' claim also gives the impression that if religion cannot successfully compete with science then religion is superfluous and undermined by science. But this is to miss the point of religion. It is to make possible a relationship with a divine Other which deepens through worship and prayer, and not to offer a competing explanation to scientific theories concerning facts about the universe and life.

But that is not to deny, however, that belief in God has in certain historical circumstances functioned as a scientific hypothesis. Almost everybody up to the second half of the nineteenth century accepted the idea that a cosmic intelligent designer created the existing biological species analogously to how we create artifacts (the artifact designer theory). Dawkins is right that evolutionary theory has undermined this religious belief but wrong in pretending that it is, therefore, no longer possible to be an intellectually fulfilled religious believer. This conclusion, however, is plausible only if evolutionary theory undermines not merely the religious belief, 'God has created the different species analogously to how we create artifacts', but also the more basic religious belief that 'God has created the world'. But we have seen that it is not the case that evolutionary theory undermines the religious belief that God has created the world or the universe. We have also noted that whereas the fortunes of naturalism as a form of intellectual belief depend upon the fortunes of the theory of evolution, the future of traditional religion does not depend upon the fortunes of the artifact designer theory since the belief that God has created the world is compatible with evolutionary theory.

Likewise, the arguments given by Dawkins, Gould, Porvine, Wilson and others that evolutionary theory alone undermines (a) the religious belief that a powerful and benevolent God could have created complex organisms by means of evolution and (b) the religious belief that there is a purpose or meaning to the existence of the universe and to human life in particular, fail in that these claims depend for their plausibility on extra-scientific premises. What is true is that scientific theories, such as evolutionary theory, can *in conjunction* with extra-scientific or philosophical claims undermine such a religious belief. Note, however, that to the extent one thinks that this is possible, one also questions the plausibility of T1. This is so because such extra-scientific premises are, of course, not species of scientific knowledge. But to have force they must be considered to be true, that is, someone (or hopefully many) either knows them to be true or is at least rationally entitled to believe them to be true. So if we take these extra-scientific claims seriously, by doing so we also question the truth of T1.

The last question we have considered is whether science can provide an alternative to traditional religion (that is, project C). The idea is that we could and should replace traditional religion with scientific naturalism and become science believers. The essential problem with this new religion or view of life is not whether it can fulfil the conditions for being a religion or view of life, but whether it is science or a proper part of science. The difficulty is that scientific naturalism, in contrast to philosophical naturalism, pretends to be science or to be a necessary presupposition of science or to be directly implied by science. But it is something we have good reason to reject. We have seen that to work properly science needs to presuppose naturalism only in the methodological sense. Theories in chemistry, biology and physics can be true, and predict and explain events in the world, whether or not it is the case that matter or physical nature alone is ultimately real. Nor is it the case that science directly implies metaphysical naturalism in general or that science can show, for instance, that humans are nothing but matter in motion. Science cannot exclude the possibility that there are dimensions of reality that are neither describable in scientific language nor accessible to scientific explanations. Crucial for a religion is also its capacity to tell its adherents what they ought to value, how they ought to behave towards other people and what interests they should develop and focus their attention on in life in order to live a morally good and meaningful life. But we have seen that this is not something science in general or evolutionary biology in particular can do. So scientific naturalism can become a religion, a view of life or mythology only if its advocates add certain extra-scientific claims to the scientific theories they base their beliefs on. It is the conflation of these elements that gives the false impression that science can be one's religion and in particular that evolutionary theory can be the best myth we will ever have. Scientific naturalists, thus, face a dilemma: *either* scientific naturalists maintain that what they are doing is science but then have to give up their missionary activities or their naturalism and become merely scientists *or* scientific naturalists keep their naturalism but then have to admit that they are not doing science anymore. We ought, therefore, not to become science believers.

Nevertheless, it is true that science has transformed our intellectual landscape. It is difficult even to imagine what our intellectual life would be without it. Hilary Putnam writes, 'and if what impressed the Few about science from the start was its stunning intellectual success, there is no doubt that what has impressed the Many is its overwhelming material and technological success. We are impressed by this even when it threatens our very lives' (Putnam 1990: 176). It is, therefore, not surprising that we encounter scientists who believe that there are no real limits to the competence of science or no limits to what can be achieved in the name of science, scientists who endorse Scientism or scientific expansionism. Putting it crudely, many scientists as well as others look to science in the hope that it will bring them salvation. Science in the life of many people fulfils a religious or quasi-religious or mythological function.

This is well worth having in mind because scientific expansionists, of course, want to give the impression that the motive of their Scientism or expansionism is merely the previous success of the scientific enterprise, that is, of the success of the *natural* sciences. Wilson, for instance, writes,

> it is astonishing that the study of ethics has advanced so little since the nineteenth century. The most distinguishing and vital qualities of the human species remain a blank space on the scientific map. I doubt that discussions of ethics should rest upon the freestanding assumptions of contemporary philosophers who have evidently never given thought to the evolutionary origin and material functioning of the human brain. In no other domain of the humanities is a union with the natural sciences more urgently needed.
>
> (Wilson 1998: 62).

So if only the social sciences, the humanities and theology could be transformed into natural science, these areas of human life could much be improved and the stunning intellectual success of (natural) science can continue. This is one motive behind Scientism.

But another less frequently mentioned motive is the attraction of being able to reach salvation through science and therefore to dispense forever with traditional religion. As Mary Midgley, Alvin Plantinga and others point out, we all need a big picture, something that brings things together and makes sense of the whole. Midgley writes that we have a choice of what myths or what visions we can use to help us understand the world, but 'we do not have a choice of understanding it without using any myths or visions at all' (Midgley 1992: 13). Myth in this sense means 'a shared way of understanding ourselves at the deep level of religion, a deep interpretation of ourselves to ourselves, a way of telling us why we are here, where we come from, and where we are going' (Plantinga 1991: 17). This is a function evolutionary theory in particular is well suited to satisfy. As we have seen, Wilson believes that 'the evolutionary epic is probably the best myth we will ever have' and Dawkins maintains that 'Darwin made it possible to be "an intellectually fulfilled atheist"' (Wilson 1978: 201 and Dawkins 1986: 6).

That there is such a 'religious' motive behind Scientism has also been pointed out by some sceptics. Thomas Nagel wonders 'why so many people welcome Darwinist imperialism' and thinks that it has to do with the 'fear of religion', that is, one wanting atheism or naturalism to be true and hoping that there is no God, but being made uneasy by the fact that many intelligent and well-informed people are religious believers (Nagel 1997: 130, 133). This fear of religion or 'cosmic authority problem',

> is not a rare condition and [Nagel's guess is that it is] responsible for much of the Scientism and reductionism of our time. One of the tendencies it supports is the ludicrous overuse of evolutionary biology to explain everything about life, including everything about the human mind. Darwin enabled modern secular culture to heave a great collective sigh of relief, by apparently providing a way to eliminate purpose, meaning, and design as fundamental features of the world. Instead they become epiphenomena.
>
> (Nagel 1997: 131)

What exactly is the rationale behind Scientism, is, of course, hard to determine, but both the 'success motive' and the 'religious motive' are well worth keeping in mind. Nevertheless, the closer the issues are to what is distinctively human, the deeper we can expect the ideological or religious involvement to be. A lesson to be learned from this study is, therefore, that the public has to be more suspicious about what is claimed in the name of science, and scientists themselves need to be less naive about the impact of their own ideological beliefs or value commitments on their scientific theorizing. What is called science can be far from an objective and dispassionate attempt to figure out the truth entirely independent of theism and naturalism, or of political and moral convictions. It is important that both the general public and scientists themselves become aware of this.

It is the conflation of these elements that gives the false impression that science can be one's religion and replace both traditional religion and ethics. The truly scientific mind must instead be conscious of the limitation of the scientific enterprise, and also allow forms of truth and knowledge which lie beyond the scope of the sciences. Some truths can only be approached by personal commitment and involvement and sometimes when our lives need to be redeemed or transformed, what is crucial is not predictions, experiment and control but the disclosure of unanticipated new meanings where old ones have been shattered. In fulfilling this task, the task of making reality existentially intelligible, it is religion and not science that has proved to be useful; and there is, as we have seen, no reason to change this belief – despite the claims of some contemporary evolutionary biologists. This study, thus, provides support for the idea that 'acceptance of an evolutionary account of the origin of human intelligence leaves ample scope for humans to develop meaning, values, and purpose (including religious meaning, values, and purpose) on a cultural level' (van Huyssteen 2000: 49).

Bibliography

Alexander, Richard D. (1987) *The Biology of Moral Systems*. New York: Aldine De Gruyter.

Alston, William P. (1991) *Perceiving God*. Ithaca, NY: Cornell University Press.

Ashmore, Robert B. (1987) *Building a Moral System*. Englewood Cliffs, NJ: Prentice-Hall.

Ayala, Francisco J. (1987) 'The Biological Roots of Morality', *Biology and Philosophy*, 2 (3).

Bannister, Robert C. (1987) *Sociology and Scientism*. Chapel Hill: University of North Carolina Press.

Barbour, Ian (1990) *Religion in an Age of Science*. New York: Harper & Row.

Basinger, David (1996) *The Case for Freewill Theism*. Downers Grove, Ill.: InterVarsity Press.

Beauchamp, Tom L. (1982) *Philosophical Ethics*. New York: McGraw-Hill.

Bleicher, Josef (1982) *The Hermeneutic Imagination: Outline of a Positive Critique of Scientism and Sociology*. London: Routledge & Kegan Paul.

Braithwaite, R. B. (1971) 'An Empiricist's View of the Nature of Religious Belief', in Basil Mitchell (ed.) *The Philosophy of Religion*. London: Oxford University Press.

Brooke, John Hedley (1991) *Science and Religion*. Cambridge: Cambridge University Press.

Bråkenhielm, Carl Reinhold and Mats G. Hansson (1995) *Livets grundmönster och mångfald* [The Basic Patterns of Life and its Diversity]. Stockholm: Liber Utbildning.

Callicott, J. Baird (1989) 'Introduction: The Real Work', *In Defense of the Land Ethic*. Albany, NY: SUNY Press.

Cannon, Dale (1996) *Six Ways of Being Religious*. Belmont, NY: Wadsworth.

Carnap, Rudolf (1967) *The Logical Structure of the World*. Berkeley: University of California Press.

Chalmers, Alan (1990) *Science and its Fabrication*. Milton Keynes: Open University Press.

Churchland, Paul (1979) *Scientific Realism and the Plasticity of Mind*. Cambridge: Cambridge University Press.

Clarke, Peter B. and Peter Byrne (1993) *Religion Defined and Explained*. New York: St. Martin's Press.

Clayton, Philip (1997) *God and Contemporary Science*. Grand Rapids, Mich.: Eerdmans.

Crick, Francis (1966) *Of Molecules and Men*. Seattle: University of Washington Press.

Crick, Francis (1994) *The Astonishing Hypothesis: The Scientific Search for the Soul*. New York: Charles Scribner's Sons.

Dawkins, Richard (1986) *The Blind Watchmaker*. New York: W.W. Norton.

Dawkins, Richard (1989) *The Selfish Gene*. (2nd edn. [1976]) Oxford: Oxford University Press.

Dawkins, Richard (1992) article in *The Independent*, 16 April.

Dawkins, Richard (1995a) *River Out of Eden*. New York: Basic Books.

Dawkins, Richard (1995b) 'A Reply to Poole', *Science and Christian Belief*, 7 (1).

Dennett, Daniel C. (1991) *Consciousness Explained*. New York: Little, Brown.

Dennett, Daniel C. (1995) *Darwin's Dangerous Idea*. Harmondsworth: Penguin.

Fagerström, Torbjörn (1994) 'Har Naturen Någon Moral?' ['Has Nature any Morality?'], in Anders Jeffner and Nils Uddenberg (eds.), *Biologi och livsåskådning*. Stockholm: Natur och Kultur.

Gorski , Philip S. (1990) 'Scientism, Interpretation, and Criticism', *Zygon*, 25 (3).

Gould, Stephen Jay (1977) *Ever Since Darwin*. Harmondsworth: Penguin.

Gould, Stephen Jay (1983) 'Extemporaneous Comments on Evolutionary Hope and Realities', in Charles L. Hamrum (ed.) *Darwin's Legacy, Nobel Conference XVIII*. San Franscisco: Harper & Row.

Graham, Loren R. (1981) *Between Science and Values*. New York: Columbia University Press.

Haack, Susan (1993) *Evidence and Inquiry*. Oxford: Blackwell.

Hagin, Kenneth E. (1978) *The Human Spirit*, vol. II. Tulsa, Okla.: Faith Library Publications.

Hasker, William (1989) *God, Time and Knowledge*. Ithaca, NY: Cornell University Press.

Haught, John F. (1995) *Science and Religion*. New York: Paulist Press.

Haught, John F. (2000) *God After Darwin*. Boulder, Col.: Westview Press.

Hawking, Stephen W. (1988) *A Brief History of Time*. London: Bantam Press.

Heidegger, Martin (1962) *Being and Time*. Oxford: Blackwell.

Herrmann, Eberhard (1995) *Scientific Theory and Religious Belief: An Essay on the Rationality of Views of Life*. Kampen: Kok Pharos.

Hick, John. (1966) *Evil and the God of Love*. London: Macmillan.

Holmberg, Martin (1994) *Narrative, Transcendence and Meaning*. Uppsala: Almqvist & Wiksell.

Howard-Snyder, Daniel (ed.) (1996) *The Evidential Argument from Evil*. Bloomington: Indiana University Press.

Huyssteen van, Wentzel (2000) 'Evolution: The Key to Knowledge of God?', in Russell Stannard (ed.) *God for the 21st Century*. Philadelphia, Pa.: Templeton Foundation Press.

Irvine, S. and A. Ponton (1988) *A Green Manifesto: Policies for a Green Future*. London: Macdonald Optima.

Jeffner, Anders (1978) *Livsåskådningsforskning* [Life-View Studies]. Research report, Uppsala University.

Kierkegaard, Søren (1974) *Concluding Unscientific Postscript*. Princeton, NJ.: Princeton University Press.

Kitcher, Philip (1986) *Vaulting Ambition*. Cambridge, Mass.: MIT Press.

Lewontin, R. C. (1997) 'Billions and Billions of Demons', (a review of Carl Sagan's book *The Demon-Haunted World*), *The New York Review of Books*, 9 January.

Lindberg, David C. and Ronald L. Numbers (eds) (1986) *God and Nature*. Berkeley: University of California Press.

McMullin, Ernan (1991) 'Plantinga's Defense of Special Creation', *Christian Scholar's Review*, 21 (1).

Midgley, Mary (1992) *Science as Salvation*. London: Routledge.

Moser, Paul K. and J. D. Trout (eds) (1995) *Contemporary Materialism*. London: Routledge.

Murphy, Nancey (1993) 'Evidence of Design in the Fine-Tuning of the Universe', in Robert John Russell, Nancey Murphy and C. J. Isham (eds) *Quantum Cosmology and the Laws of Nature*. Notre Dame: University of Notre Dame Press.

Naess, Arne (1989) *Ecology, Community and Lifestyle*. Cambridge: Cambridge University Press.

Nagel, Thomas (1979) *Mortal Questions*. Cambridge: Cambridge University Press.

Nagel, Thomas (1997) *The Last Word*. New York: Oxford University Press.

Nielsen, Kai (1999) 'Naturalistic Explanations of Religion', *Studies in Religion*. 26 (4).

O'Hear, Anthony (1997) *Beyond Evolution*. Oxford: Oxford University Press.

Peacocke, Arthur (1993) *Theology for a Scientific Age*. Minneapolis: Fortress Press.

Peterson, Michael, William Hasker, Bruce Reichenbach and David Basinger (1991) *Reason and Religious Belief*. Oxford: Oxford University Press.

Plantinga, Alvin (1983) 'Reason and Belief in God', in Alvin Plantinga and Nicholas Wolterstorff (eds) *Faith and Rationality*. Notre Dame: University of Notre Dame Press.

Plantinga, Alvin (1991) 'When Faith and Reason Clash', *Christian Scholar's Review*, 21 (1).

Plantinga, Alvin (1996) 'Science: Augustinian or Duhemian?', *Faith and Philosophy*, 13 (3).

Plantinga, Alvin (2000) *Warranted Christian Belief*. Oxford: Oxford University Press.

Polkinghorne, John (1996) *Beyond Science*. Cambridge: Cambridge University Press.

Polkinghorne, John (1998) *Belief in God in an Age of Science*. New Haven, Conn.: Yale University Press.

Poole, Michael, W. (1994) 'A Critique of Aspects of the Philosophy and Theology of Richard Dawkins', *Science and Christian Belief*. 6 (1).

Popper, Karl R. and John C. Eccles (1977) *The Self and Its Brain*. London: Springer International.

Post, John F. (1991) *Metaphysics*. New York: Paragon House.

Provine, William (1988) 'Evolution and the Foundation of Ethics', *MBL Science*, 3.

Putnam, Hilary (1981) *Reason, Truth and History*. Cambridge: Cambridge University Press.

Radnitzky, Gerard (1978) 'The Boundaries of Science and Technology', in Proceedings of the Sixth Internal Conference on the Unity of the Sciences, *The Search for Absolute Values in a Changing World*, vol II. New York: The International Cultural Foundation Press.

Regan, Tom (1983) *The Case for Animal Rights*. Berkeley: University of California Press.

Rescher, Nicholas (1984) *The Limits of Science*. Berkeley: University of California Press.

Rolston, Holmes (1994) *Conserving Natural Value*. New York: Columbia University Press.

Rolston, Holmes (1999) *Genes, Genesis and God*. Cambridge: Cambridge University Press.

Ruse, Michael and Edward O. Wilson (1986) 'Moral Philosophy as Applied Science', *Philosophy*, 61.

Ruse, Michael and Edward O. Wilson (1993) 'The Evolution of Ethics', in James E. Huchingson (ed.) *Religion and the Natural Sciences: The Range of Engagement*. San Diego, Calif.: Harcourt Brace.

Ruse, Michael (1985) *Sociobiology: Sense or Nonsense?* Dordrecht: D. Reidel.

Ruse, Michael (1998) *Taking Darwin Seriously.* (2nd edn.) [1986]. Oxford: Blackwell.

Russell, Bertrand (1957) *Why I am not a Christian*. London: Unwin Paperbacks.

Sagan, Carl (1980) *Cosmos*. New York: Ballantine Books.

Sagan, Carl (1997) *The Demon-Haunted World: Science as a Candle in the Dark*. London: Headline.

Schoeck, Helmut and James W. Wiggins (eds) (1960) *Scientism and Values*. Princeton, NJ: D. van Nostrand.

Settle, Tom (1995) 'You Can't Have Science as Your Religion!', in I. C. Jarvie and Nathaniel Laor (eds.) *Critical Rationalism, Metaphysics and Science*, vol. 1. Dordrecht: Kluwer Academic.

Simon, Herbert (1990) 'A Mechanism for Social Selection and Successful Altruism', *Science*, 250.

Simpson, George Gaylord (1967) *The Meaning of Evolution*. (rev. edn.) New Haven, Conn.: Yale University Press.

Singer, Peter (1981) *The Expanding Circle*. Oxford: Clarendon Press.

Singer, Peter (1993) 'Ethics and Sociobiology', in James E. Huchingson (ed.) *Religion and the Natural Sciences: The Range of Engagement*, San Diego, Calif.: Harcourt Brace.

Smart, J. J. C. (1996) 'Atheism and Theism', in J. J. C. Smart and J. Haldane, *Atheism and Theism*. Oxford: Blackwell.

Sorell, Tom (1991) *Scientism*. London: Routledge.

Stenmark, Mikael (1995a) *Rationality in Science, Religion, and Everyday Life: A Critical Evaluation of Four Models of Rationality*. Notre Dame: University of Notre Dame Press.

Stenmark, Mikael (1995b) 'Guds relevans för livets mening', ['God and the Meaning of Life'] *Svensk Teologisk Kvartalskrift*, 71 (1).

Stenmark, Mikael (1997) 'An Unfinished Debate: What are the Aims of Science and Religion?', *Zygon: Journal of Religion and Science*, 32 (4).

Stenmark, Mikael (1999) 'Rationality and Religious Commitment', *Studia Theologica*, 53.

Stevenson, Leslie and Henry Byerly (1995) *The Many Faces of Science*. Boulder, Col.: Westview Press.

Stich, Stephen (1985) *From Folk Psychology to Cognitive Science*. Boston, Mass.: MIT Press.

Stove, David (1994) 'So You Think You Are a Darwinian?', *Philosophy*, 69.

Swinburne, Richard (1979) *The Existence of God*. Oxford: Oxford University Press.

Swinburne, Richard (1983) 'Mackie, Induction, and God', *Religious Studies*, 19.

Swinburne, Richard (1996) *Is there a God?* Oxford: Oxford University Press.

Trigg, Roger (1993) *Rationality and Science*. Oxford: Blackwell.

Trivers, Robert L. (1971) 'The Evolution of Reciprocal Altruism', *Quarterly Review of Biology*, 46.

Ward, Keith (1996) *God, Change and Necessity*. Oxford: Oneworld Publications.

Williams, George C. (1988) 'Huxley's Evolution and Ethics in Sociobiological Perspective', *Zygon*, 23.

Wilson, Edward O. (1975) *Sociobiology*. Cambridge, Mass.: Harvard University Press.

Wilson, Edward O. (1978). *On Human Nature*. Cambridge, Mass.: Harvard University Press.

Wilson, Edward O. (1994) *Naturalist*. Washington, DC: Island Press.

Wilson, Edward O. (1998) 'The Biological Basis of Morality', *The Atlantic Monthly*, April.

Wykstra, Stephen, J. (1990) 'The Humean Obstacle to Evidential Arguments from Suffering', in M. M. Adams and R. M. Adams (eds) *The Problem of Evil*. Oxford: Oxford University Press.

Zagzebski, Linda (1991) *The Dilemma of Freedom and Foreknowledge*. Oxford: Oxford University Press.

Index